W0115088

Lecture Notes in Physics

For information about Vols. 1–172, please contact your bookseller or Springer-Verlag.

Vol. 173: Stochastic Processes in Quantum Theory and Statistical Physics. Proceedings, 1981. Edited by S. Albeverio, Ph. Combe, and M. Sirugue-Collin. VIII, 337 pages. 1982.

Vol. 174: A. Kadić, D.G.B. Edelen, A Gauge Theory of Dislocations and Disclinations. VII, 290 pages. 1983.

Vol. 175: Defect Complexes in Semiconductor Structures. Proceedings, 1982. Edited by J. Giber, F. Beleznay, J. C. Szép, and J. László. VI, 308 pages. 1983.

Vol. 176: Gauge Theory and Gravitation. Proceedings, 1982. Edited by K. Kikkawa, N. Nakanishi, and H. Nariai. X, 316 pages. 1983.

Vol. 177: Application of High Magnetic Fields in Semiconductor Physics. Proceedings, 1982. Edited by G. Landwehr. XII, 552 pages. 1983.

Vol. 178: Detectors in Heavy-Ion Reactions. Proceedings, 1982. Edited by W. von Oertzen. VIII, 258 pages. 1983.

Vol. 179: Dynamical Systems and Chaos. Proceedings, 1982. Edited by L. Garrido. XIV, 298 pages. 1983.

Vol. 180: Group Theoretical Methods in Physics. Proceedings, 1982. Edited by M. Serdaroğlu and E. İnönü. XI, 569 pages. 1983.

Vol. 181: Gauge Theories of the Eighties. Proceedings, 1982. Edited by R. Raitio and J. Lindfors. V, 644 pages. 1983.

Vol. 182: Laser Physics. Proceedings, 1983. Edited by J. D. Harvey and D. F. Walls. V, 263 pages. 1983.

Vol. 183: J. D. Gunton, M. Droz, Introduction to the Theory of Metastable and Unstable States. VI, 140 pages. 1983.

Vol. 184: Stochastic Processes – Formalism and Applications. Proceedings, 1982. Edited by G. S. Agarwal and S. Dattagupta. VI, 324 pages. 1983.

Vol. 185: H. N. Shirer, R. Wells, Mathematical Structure of the Singularities at the Transitions between Steady States in Hydrodynamic Systems. XI, 276 pages. 1983.

Vol. 186: Critical Phenomena. Proceedings, 1982. Edited by F.J.W. Hahne. VII, 353 pages. 1983.

Vol. 187: Density Functional Theory. Edited by J. Keller and J.L. Gázquez. V, 301 pages. 1983.

Vol. 188: A. P. Balachandran, G. Marmo, B.-S. Skagerstam, A. Stern, Gauge Symmetries and Fibre Bundles. IV, 140 pages. 1983.

Vol. 189: Nonlinear Phenomena. Proceedings, 1982. Edited by K. B. Wolf. XII, 453 pages. 1983.

Vol. 190: K. Kraus, States, Effects, and Operations. Edited by A. Böhm, J. W. Dollard and W. H. Wootters. IX, 151 pages. 1983.

Vol. 191: Photon Photon Collisions. Proceedings, 1983. Edited by Ch. Berger. V, 417 pages. 1983.

Vol. 192: Heidelberg Colloquium on Spin Glasses. Proceedings, 1983. Edited by J. L. van Hemmen and I. Morgenstern. VII, 356 pages. 1983.

Vol. 193: Cool Stars, Stellar Systems, and the Sun. Proceedings, 1983. Edited by S. L. Balliunas and L. Hartmann. VII, 364 pages. 1984.

Vol. 194: P. Pascual, R. Tarrach, QCD: Renormalization for the Practitioner. V, 277 pages. 1984.

Vol. 195: Trends and Applications of Pure Mathematics to Mechanics. Proceedings, 1983. Edited by P. G. Ciarlet and M. Roseau. V, 422 pages. 1984.

Vol. 196: WOPPLOT 83. Parallel Processing: Logic, Organization and Technology. Proceedings, 1983. Edited by J. Becker and I. Eisele. V, 189 pages. 1984.

Vol. 197: Quarks and Nuclear Structure. Proceedings, 1983. Edited by K. Bleuler. VIII, 414 pages. 1984.

Vol. 198: Recent Progress in Many-Body Theories. Proceedings, 1983. Edited by H. Kümmel and M. L. Ristig. IX, 422 pages. 1984.

Vol. 199: Recent Developments in Nonequilibrium Thermodynamics. Proceedings, 1983. Edited by J. Casas-Vázquez, D. Jou and G. Lebon. XIII, 485 pages. 1984.

Vol. 200: H. D. Zeh, Die Physik der Zeitrichtung. V, 86 Seiten. 1984.

Vol. 201: Group Theoretical Methods in Physics. Proceedings, 1983. Edited by G. Denardo, G. Ghirardi and T. Weber. XXXVII, 518 pages. 1984.

Vol. 202: Asymptotic Behavior of Mass and Spacetime Geometry. Proceedings, 1983. Edited by F. J. Flaherty. VI, 213 pages. 1984.

Vol. 203: C. Marchioro, M. Pulvirenti, Vortex Methods in Two-Dimensional Fluid Dynamics. III, 137 pages. 1984.

Vol. 204: Y. Waseda, Novel Application of Anomalous (Resonance) X-Ray Scattering for Structural Characterization of Disordered Materials. VI, 183 pages. 1984.

Vol. 205: Solutions of Einstein's Equations: Techniques and Results. Proceedings, 1983. Edited by C. Hoenselaers and W. Dietz. VI, 439 pages. 1984.

Vol. 206: Static Critical Phenomena in Inhomogeneous Systems. Edited by A. Pękalski and J. Sznajd. Proceedings, 1984. VIII, 358 pages. 1984.

Vol. 207: S. W. Koch, Dynamics of First-Order Phase Transitions in Equilibrium and Nonequilibrium Systems. III, 148 pages. 1984.

Vol. 208: Supersymmetry and Supergravity/Nonperturbative QCD. Proceedings, 1984. Edited by P. Roy and V. Singh. V, 389 pages. 1984.

Vol. 209: Mathematical and Computational Methods in Nuclear Physics. Proceedings, 1983. Edited by J. S. Dehesa, J. M. G. Gomez and A. Polls. V, 276 pages. 1984.

Vol. 210: Cellular Structures in Instabilities. Proceedings, 1983. Edited by J. E. Wesfreid and S. Zaleski. VI, 389 pages. 1984.

Vol. 211: Resonances – Models and Phenomena. Proceedings, 1984. Edited by S. Albeverio, L. S. Ferreira and L. Streit. VI, 369 pages. 1984.

Vol. 212: Gravitation, Geometry and Relativistic Physics. Proceedings, 1984. Edited by Laboratoire "Gravitation et Cosmologie Relativistes", Université Pierre et Marie Curie et C.N.R.S., Institut Henri Poincaré, Paris. VI, 336 pages. 1984.

Vol. 213: Forward Electron Ejection in Ion Collisions. Proceedings, 1984. Edited by K. O. Groeneveld, W. Meckbach and I. A. Sellin. VII, 165 pages. 1984.

Lecture Notes in Physics

237

Nearby Molecular Clouds

Proceedings of a Specialized Colloquium of the
Eighth IAU European Regional Astronomy Meeting
Toulouse, September 17–21, 1984

Edited by G. Serra

Springer-Verlag Berlin Heidelberg GmbH

Editor

Guy Serra
CESR-CNRS/UPS
9 avenue du Colonel Roche, F-31029 Toulouse Cédex

ISBN 978-3-540-15991-9 ISBN 978-3-540-39696-3 (eBook)
DOI 10.1007/978-3-540-39696-3

Originally published by Springer-Verlag Berlin Heidelberg New York Tokyo in 1985

COLLOQUE

NUAGES MOLECULAIRES PROCHES

Toulouse, 17-21 Septembre 1984

HUITIEME ASSEMBLEE
REGIONALE EUROPEENNE

Union Astronomique Internationale
Société Européenne de Physique

COLLOQUIUM

NEARBY MOLECULAR CLOUDS

Toulouse, September 17-21, 1984

EIGHTH EUROPEAN
REGIONAL MEETING

International Astronomical Union
European Physical Society

COMITE SCIENTIFIQUE D'ORGANISATION

SCIENTIFIC ORGANIZING COMMITTEE

L. BLITZ

G.B. SHOLOMITSKII

T. DE GRAAUW

E. FALGARONE

T. MONTMERLE

P.C. MYERS

A. NATTA

J.L. PUGET

G. SERRA (Chairman)

T. WILSON

FOREWORD

The colloquium "Nearby Molecular Clouds" was one of the three specialized colloquia included in the Eighth European Regional Astronomy Meeting of the International Astronomical Union and the European Physical Society. It was held at the CESR-CNRS/UPS in Toulouse (France) from September 17 to 21, 1984. More than 70 scientists (mainly from Europe, but also from Mexico, the United States of America and the USSR) attended to report and discuss recent advances in the study of the interstellar Medium (I.M.) and Star Formation (S.F.).

As mentioned by several scientists taking part in this meeting, the "Nearby Molecular Clouds" (N.M.C.) may provide specific critical information about I.M. and S.F. The observations of N.M.C. take advantage of the higher spatial resolution and the better sensitivity to weak sources over a larger fraction of the electromagnetic spectrum.

Many recent infrared (IRAS) and millimetric observations make these topics of particular interest now. These new data make possible intercomparisons between nearby objects, which are very useful in understanding general trends in the cloud evolution and in the physics of the relations between (1) the stars already formed inside the cloud, (2) the gas and (3) the dust. It is now clear, also, that the increase of observational results in this field allows new theoretical work devoted to cloud physics and S.F. processes.

This colloquium was aimed at a better understanding of all these topics. New results were presented in the papers (edited here) and very interesting discussions occurred during the colloquium. Unfortunately, it was not possible to publish the proceedings of the discussions because the questions/answers written and sent to me by the authors were very inhomogeneous and incomplete.

I would like to thank everybody who participated in the colloquium and the many people who helped with the organisation; in particular, E. FALGARONE and T. MONTMERLE, who played extensive roles in the scientific organisation of the colloquium. I wish also to acknowledge the assistance afforded by CESR-CNRS/UPS people in the technical organisation, in particular, its director, Pr. G. VEDRENNE, its administrator, J.P. CHAMPAGNAC and E. DELMAS and E. CASSOUX who afforded the secretarial and translation support.

I hope that the present proceedings will prove useful not only to those who were present at this colloquium but also to all the scientists who were not able to attend the meeting.

Toulouse, July 1985

Guy Serra

AVANT-PROPOS

Le Colloque "Nuages Moléculaires Proches" a été l'un des trois colloques spécialisés réalisés dans le cadre de la Huitième Assemblée Régionale Européenne de l'Union Astronomique Internationale et de la Société Européenne de Physique.

Il s'est tenu au CESR-CNRS/UPS à Toulouse (FRANCE) du 17 au 21 septembre 1984. Plus de 70 scientifiques, principalement européens mais aussi d'autres pays: Mexique, Etats-Unis d'Amérique et URSS ont participé à la présentation et à la discussion de récentes avancées dans l'étude du Milieu Interstellaire, (M.I.) et de la Formation des Etoiles (F.E.).

Comme cela a été mentionné par plusieurs chercheurs participant à cette réunion, les Nuages Moléculaires Proches (N.M.P.) permettent un accès spécifique à des informations critiques relatives au M.I. et à la F.E. L'observation des N.M.P. présente l'avantage d'un gain en résolution spatiale et en sensibilité aux sources faibles sur une large portion du spectre électromagnétique.

De nombreuses observations récentes dans le domaine des émissions infrarouge et millimétrique donnent aujourd'hui un regain d'intérêt à ces sujets. Ces nouveaux résultats observationnels rendent possibles des intercomparaisons entre objets proches, intercomparaisons très intéressantes pour dégager les effets dominants qui déterminent l'évolution des Nuages Interstellaires et pour mieux comprendre la physique des relations entre: 1) les étoiles déjà formées dans le Nuage, 2) le gaz et, 3) les poussières. Il est maintenant clair, aussi, que l'accumulation d'observations nouvelles dans ce domaine, permettent des travaux théoriques inédits consacrés à la physique des Nuages et aux processus de formation d'Etoiles.

Le but de ce Colloque a été d'améliorer notre compréhension dans ces champs d'investigation. De nouveaux résultats ont été présentés dans les articles édités dans cet ouvrage; de très intéressantes discussions ont eu lieu durant ce Colloque. Malheureusement, il n'a pas été possible de publier les actes de ces discussions en raison des caractères inhomogène et incomplet des textes écrits pour les questions et réponses qui me sont parvenus.

Je voudrais remercier tous les scientifiques ayant participé à ce Colloque et toutes les personnes qui ont contribué à son organisation; en particulier, E. FALGARONE et T. MONTMERLE qui ont pris une très large part au travail d'orgasation scientifique. Je souhaite aussi remercier pour leur assistance dans l'organisation technique du Colloque les membres du CESR-CNRS/UPS, en particulier: son Directeur Pr G. VEDRENNE, son Administrateur J.P. CHAMPAGNAC et E. DELMAS et E. CASSOUX qui ont assuré les tâches de secrétariat et de traduction.

Je souhaite que ces actes soient utiles non seulement aux participants du Colloque, mais aussi à de nombreux scientifiques qui n'ont pas pu y participer.

Toulouse, juillet 1985 — Guy Serra

TABLE OF CONTENTS

PART I: Nearby Molecular Clouds as Part of Larger Structure

PART II: Compared Internal Structure of Nearby Molecular Clouds

PART III: Star Formation in Nearby Molecular Clouds

PART IV: Miscellaneous Related Topics

PART V: Stellar Content of Nearby Molecular Clouds

The following two papers were not submitted for these proceedings:

L. BLITZ and L. MANGANI:
The Nearest Molecular Clouds

J. CERNICHARO:
Large Scale Properties of Dark Clouds: The Taurus, Auriga Persens Complex

PART I

NEARBY MOLECULAR CLOUDS AS PART OF LARGER STRUCTURE

A SYNTHETIC VIEW AT LARGE SCALE OF LOCAL MOLECULAR CLOUDS

F. Lebrun
Service d'Astrophysique, C.E.N. Saclay
91191 Gif/Yvette CEDEX, France

ABSTRACT

Local molecular clouds can be studied in many ways at various wavelenghts from radio waves to γ rays. None of these observations leads alone to a single and unambigous interpretation. Simplifying assumptions are generarally made which limit the degree of confidence of the results obtained. As an example it is often assumed, especially in the local interstellar medium, that some quantities are constant. Among them I can list, the gas to dust ratio, $N(H_2)$/Wco ratio, the HI spin temperature, the cosmic-ray density, etc...

Synthetic studies may then help resolving some of the ambiguities resulting from these assumptions. An example is discussed based on the comparison of galaxy counts, ~ rays, U-V absorption lines, CO, star counts and reddening. Furthermore, these observations are also very often complementary. The 21 cm sky surveys provide an accurate and nearly fully sampled map of atomic hydrogen, but does not depict the H_2 distribution. Gamma rays can trace both species but are not very accurate. The reddening measurements cannot provide a fully sample sky survey and are also not very accurate but are probably the only way of estimating cloud distances in the solar neighbourhood (d < 500 pc), which are essential for estimating sizes and masses.

WARNING: The present author being involved in γ-ray astronomy, his "synthetic" view of local molecular clouds is biased towards a rough estimate of the total gas column density.

1 - INTRODUCTION

In the 1930's interstellar clouds were exclusively nearby dark clouds and remained such until the beginning of radioastronomy. Large scale studies were essentially attempts to catalog such objects brought into evidence by the variations of the star number density (Lundmark 1926, Barnard 1927). Hubble (1933) provided a more uniform view of the dust content of the interstellar medium (ISM) with the use of galaxy counts.

That dust clouds were also gas cloud was mainly established by a comparison of these galaxy counts with an early 21 cm survey of the galactic anticenter (Lilley 1955).

Subsequently, there was some debate about the degree of correlation between gas and dust in the ISM (Garzoli and Varsavsky 1966, Heiles 1967, Sturch 1969, Wesselius and Sancisi 1971, Savage and Jenkins 1972, Knapp and Kerr 1974, Jenkins and Savage 1974, Ryter, Cesarsky and Audouze 1975). This debate ended with the publication of the Copernicus satellite measurements of U-V absorption lines towards 100 O-B stars. These first measures of molecular hydrogen column densities ($N(H_2)$) showed that there is a good linear correlation between gas and dust, as long as the gas is considered in both atomic and molecular form (Bohlin et al. 1978).

Following the discovery of CO in the Orion nebula (Wilson, Jefferts and Penzias 1970), dark clouds were found to be CO emitters (Penzias et al. 1972). Later on, Dickman (1978) found that the ratio between ^{13}CO column density and the optical absorption is roughly constant.

By about the same time preliminary results of the COS-B γ-ray mission permitted the identification of the largest dark nebula of the Lynds catalog (1962), representing most of the Aquila Rift, with a broad γ-ray excess (Lebrun and Paul 1978). The ρ Ophiuchi cloud has been barely resolved by the γ-ray telescope aboard COS-B (Swanenburg et al. 1981) and the Orion cloud complex has been the subject of detailed comparaisons of Apg, HI, CO and γ rays (Caraveo et al. 1980 and 1981, Bloemen et al. 1984).

2 - AVAILABLE LOCAL GAS-TRACER DATA

At present time, local molecular clouds appear at large scale as: optical absorbers, 21 cm and 2.6 mm emitters, U-V lines absorbers and γ-ray emitters. They are undoubtly infra-red emitters and future synthetic studies using data from the IRAS mission are very much awaited. Soft X rays can also improve our knowledge of the local ISM and recent comparisons with HI looks very promising (Mac Cammon et al. 1983). However, soft X rays reveal essentially a phase of the ISM more diffuse than that of molecular clouds.

In table 1 are listed the global properties of the local molecular-cloud tracers available at large scale. Each of them, from the radio to the γ-ray domain, will be briefly discussed in the following.

2.a 21 cm

The entire celestial sphere is covered by four 21 cm emission line surveys (Weaver and Williams 1973, Heiles and Habing 1974, Heiles and Cleary 1978, Strong et al. 1982). However the derivation of column densities require the knowledge of the distribution of the HI spin temperature (Ts). This temperature can be presently estimated only in the direction of a few hundred extragalactic sources observed in absorption at 21 cm (Crovisier et al. 1978, Dickey et al. 1978, Crovisier et al. 1980). However for large scale studies, column densities are derived on the basis of the emission measurements alone, assuming a uniform spin temperature. Furthermore, the emission in the far side lobes of the antennas may not have been completely removed by the corrections applied. So that the uncertainties associated with the 21 cm line for measuring the HI column densities are essentially systematic.

2.b 2.6 mm

With a wavelength about 2 orders of magnitude lower than that of the HI transition, the CO line has allowed high resolution surveys (~1′ with the 11m NRAO antenna) with very small coverage. At the cost of a lower resolution and sensitivity a faster coverage of the sky can be achieved (Dame and Thaddeus 1984). However the present sky coverage is still rather poor (~10%). Significant improvement of this situation is expected in the coming years with the advent of SIS detectors. Here on again, the main problem is not the statistical accuracy but rather the ability of the integrated CO line intensity (Wco) to trace the H_2 column density. Although a value of about 3 10^{20} cm^{-2} K^{-1} km^{-1} s for the galactic average of the $N(H_2)$/Wco ratio (X) would probably satisfy most radioastronomers (Lebrun et al. 1983, Sanders,Solomon and Scoville 1984), it is not evident that such a value would apply to individual local clouds.

2.c Galaxy counts

The deepest extended galaxy counts are those made by Shane and Wirtanen (1967) which covers the entire sky for $\delta > -25°$ down to the 19th magnitude. There is some uncertainty about the zero level and the scale of the absorption derived from these galaxy counts (Shane and Wirtanen 1967, Heiles 1976, Burstein and Heiles 1978, Strong and Lebrun 1982, Burstein and Heiles 1982, Strong 1983) which are more important than that on the gas to dust ratio value (ρ). However, despite these problems and the rather poor statistical accuracy of this tracer, it is still valuable since its systematic uncertainties should not vary much from one region to another.

2.d Star counts

This is the pioneer tracer. Nowdays, it is essentially used in small areas (no more than a few degrees) with an angular resolution of a few arc minutes, essentially for purpose of comparison with the millimeter wave observations (Kutner 1973, Meyers 1975, Encrenaz, Falgarone and Lucas 1975, Tucker et al. 1976, Dickman 1978, Cernicharo and Bachiller 1984). These high resolution counts are generally made on the Palomar Observatory Sky Survey plates which have a magnitude limit of about 19. At large scale, only the results of a few old studies covering a few hundred square degrees and limited to the 15th or the 16th magnitude are available (Mac Cuskey 1938, Van Hoof 1969). Their statistical accuracy is better than that of galaxy counts, but the zero level of the derived absorption is much more uncertain since it strongly depends on the choice of the reference field.

TABLE 1

LOCAL GAS TRACERS

	21cm line		2.6mm line	star counts	galaxy counts	U-V lines	E(B-V)	γ rays
	em.	abs.	em.	abs.	abs.	abs.	abs.	em.
angular resolution	0.3°		1′–8′	2′–1°	~1°			~5°
statistical accuracy (%)	~1	~1	1–10	1–10	10–100	~10	~10	~10
domain of validity	~ ∞	<1 kpc	~ ∞	<1 kpc	<1 kpc	<1 kpc	<1 kpc	~ ∞
accuracy of distance estimate (%)	~100	~100	~100	~50	none	~30	~30	none
critical assumptions	Ts=Ct	Ts=Ct	X=Ct	ρ=Ct N*°=?	ρ=Ct Ng°=Ct		ρ=Ct	C-R=Ct
sky coverage (%)	100		~10	~2	~60			~40

ρ = N(H)/Apg   X = N(H_2)/Wco   C-R = cosmic-ray flux   Ts = HI spin temperature
Ng° = number of galaxies observable in a square degree free of interstellar extinction.

2.e Reddening

The large scale distribution of the interstellar reddening material in the solar neighbourhood has been the subject of recent studies (Lucke 1978, Neckel and Klare 1980, Perry and Johnston 1982). These studies are not particularly accurate since they strongly undersample the distant material but in contrast with the previously mentionned tracer, they can provide interesting distance estimates.

2.f U-V lines

Although this tracer suffers the same type of limitations as the interstellar reddening, it is the only one able to give directly $N(H_2)$. However at present time the available data are restricted to a hundred of O-B stars observed by the Copernicus satellite and can in no way give a large scale picture of the H_2 distribution in the solar neighbourhood. At lower distances (d < 100 pc), a more precise picture of the very local ISM can emerged from the U-V measurements of various absorption lines (Paresce 1984). However, as for the soft X-ray measurements, they can only probe the rather diffuse medium.

2.g Gamma rays

Diffuse galactic γ-ray emission (E ~0.1 GeV) can basically be produced by three mechanisms: i) collisions between cosmic-ray (C-R) protons and nuclei with those of the ISM produce π^0 mesons which decay in two γ photons ii) Bremsstrahlung emission of cosmic-ray electrons in the electric field of interstellar-matter atoms iii) inverse Compton interactions of cosmic-ray electrons with the ambient photon field. Of these three mecanisms, the γ-ray production of the third one is very small in the local ISM (see e.g. Bloemen 1984). The two others involve the interstellar gas as a target for cosmic rays, so that if the cosmic-ray flux is uniform in the ISM, the diffuse γ-ray intensity in a given direction is directly proportional to the interstellar-gas column density. The cosmic-ray flux varies on galactic scale (see Bloemen et al. 1984), however this variation should be insensitive in the local ISM (d < 1 kpc). The fondamental advantage of the γ radiation as a gas tracer is its ability to reveal all the gas, disregarding its physical state: atomic, molecular or ionized. Unfortunately, present γ-ray observations are not very accurate: they suffer rather poor statistics and a low angular resolution (~5°).

3 - COMPARISONS

An important remark is that the uncertainties attached to each tracer of the gas are different in nature. Synthetic studies, or comparisons of these different tracers, should then help removing some of the systematic uncertainties resulting from the simplifying assumptions associated with the use of each tracer. In the following will be presented first the general comparison of some of these tracers in the local ISM and then particular regions of interest will be discussed case by case.

3.a General

The first attempt to study the correlations between the γ-ray intensity distribution with that of other gas tracers out of the galactic plane is due to Fichtel et al. (1978).

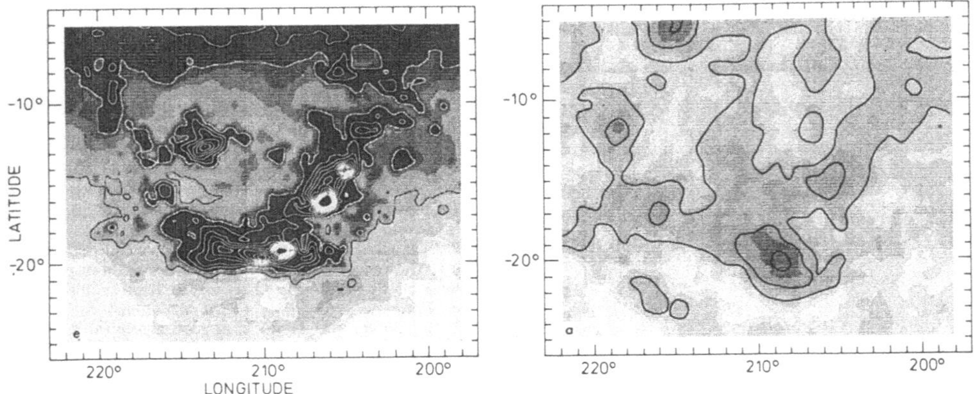

Figure 1. The Orion cloud complex. Left: map of the total gas column densities determined from Wco and N(HI) with X = 3 10^{20} cm^{-2} K^{-1} km^{-1} s, the contour interval is 2 10^{21} H atoms cm^{-2}. Right: map of the γ-ray intensity (0.1 - 5 GeV) ,the contour values are 12, 18, and 24 (10^{-5} cm^{-2} s^{-1} sr^{-1}). From Bloemen et al. (1984).

Most of the correlation observed between the γ-ray intensity and that of the 21 cm line is due to the plane structure of both emissions. Comparisons of the longitude profiles (for 10° < |b| <20°) of the 21cm emission, the galaxy counts and the γ-ray intensity (Lebrun and Paul 1983, Lebrun et al. 1982) established the two dimensional correlations between these quantities which reflects essentially the presence in these tracers of the main structural feature of the local ISM: the Gould belt. Furthermore, they showed that the correlation of the γ-ray intensity with the galaxy counts is better than with HI. This implies the presence of γ-ray emitting material linked to the dust, most likely molecular hydrogen. A map of the derived H_2 distribution was given by Strong et al. (1982). These correlations imply that on a 10° scale the gas-to-dust ratio and the C-R flux do not vary by large factors in the solar neighbourhood.

3.b Orion

So far, apart from the Galaxy itself and the Gould belt, the Orion cloud complex is the only extended gamma-ray source resolved by a γ-ray telescope (Caraveo et al. 1980). This early work has shown that the product of the gas-to-dust ratio time the γ-ray emissivity; q, was the same inside and outside the molecular cloud and equal to the local average found in the general comparison. This indicates most likely that q and ρ in the cloud are equal to their averages for the solar neighbourhood. With the progress of millimeter-wave observations, a detailed large scale map of the CO distribution is now available (Maddalena et al. 1984). These observations, together with the 21 cm data (Heiles and Habing 1974) have been compared with the γ-ray data (see fig. 1) and gave the following results (Bloemen et al. 1984): the γ-ray intensities observed can be well described by a linear combination of the two radio gas-tracers. The simplest interpretation of this result is that the gas is well depicted by both tracers, despite the high CO optical thickness, and that cosmic rays are uniform throughout the complex. Furthermore, the value found for X is similar to the galactic average. The Orion cloud is a case where all the simplifying assumptions seems to hold, at least on large scale (a few degrees).

3.c ρ Oph

The case of the γ-ray emission from the ρ Oph cloud is well known. The publication of the γ-ray flux of this source (Swanenburg et al. 1981) has triggered a number of publications (Casse and Paul 1980, Bignami and Morfill 1980, Issa, Strong and Wolfendale 1981, Morfill et al. 1981), proposing scenarii to explain an enhanced C-R density or disputing the need for such a C-R excess. Improvement of the γ-ray statistics and data processing has not solved the question since the main difficulty is to obtain reliable gas column density estimates over a reasonably large area (at least 100 degrees). The OH observations of Wouterloot (1984) may constitute a first step towards such estimates, however, the coverage of these observations is not enough uniform and extended to allow a direct comparison with the γ-ray data. A detailed study of this particularly interesting region should await the completion of a large scale (several hundred square degrees) fully sampled CO survey of the complex.

3.d Aquila Rift

This very large nearby molecular complex, containing the largest nebula catalogued by Lynds (1962) has been mapped in CO by Dame and Thaddeus (1984) during their low resolution survey of the first galactic quadrant. Its distance, as determined by the reddening measurements of Neckel and Klare (1980), is similar to that of the ρ Oph cloud : 150 pc. The optical appearence of the Aquila Rift as seen on the photograph mosaic of the Milky Way (Mt. Wilson observatory) is very similar to that in CO. The early finding of high energy γ-ray emission from the Aquila Rift region (Lebrun and Paul 1978) has been confirmed by the detailed comparison of CO, HI and γ rays in the first galactic quadrant (see fig.1 in Lebrun et al. 1983). Therefore this region appears to be perveaded by the average ambiant C-R flux and to show "normal" X and ρ values as in the case of the Orion complex.

3.e Taurus-Perseus

This complex of cloud which have been so often compared with the ρ Oph cloud is roughly in the same situation: an extended and fully sampled CO survey is missing. The very detailed observations of Baran and Thaddeus (1977) and Wouterloot (1984) do not give a sufficiently uniform coverage of the region. However as a whole this complex did not show up in the general comparison as a deviating point.

3.f Oph-Sag

This region stand out as having the lowest N(HI)/Apg ratio in Heiles (1976) study of the interelations between galaxy counts and HI. Noting that most other regions of the sky where this ratio is low correspond to well known molecular complexes, Strong and Lebrun (1982) proposed to attribute this HI deficiency to the formation of H_2. Lebrun and Paul (1983) noted that in Oph-Sag, HI is also deficient compared to the γ-ray intensity observed by SAS-2 and concluded that there exists γ-ray-emitting material linked to the dust, this material being most likely H_2. An extended (370 square degrees) and fully sampled CO survey at low resolution (0.5°) has been undertaken by Lebrun and Huang (1984) in the northern part of Oph-Sag. The CO clouds found appear to form a kind of bridge between the Aquila Rift and the ρ Oph cloud complex and follow the kinematic of the Gould belt. A comparison of these measurements with HI and the absorption traced by star counts (van Hoof 1969) indicates a normal gas to dust ratio and an X value lower ($X \sim 10^{20}$ cm^{-2} K^{-1} km^{-1} s) than that derived for the entire first galactic quadrant and in the Orion complex. The column densities predicted with this calibration based on star counts, are about 50% lower than those based on galaxy counts. The origin of this discrepancy must lie in the calibration of the absorption measurements. It could be either an underestimate of the absorption by the star counts and/or an overestimate of the absorption by the galaxy counts. Another interesting effect is that the γ-ray intensity (70 - 5000 MeV) observed by the COS-B satellite is higher than that predicted from the galaxy counts (Strong et al. 1982). This latter effect cannot be due to an error on the scale or on the zero level of the relation between galaxy counts and γ rays since both are determined by a linear fit in the general comparison. It indicates a deviation of Oph-Sag from the general trend.

4 The simplifying assumptions in question

In Oph-Sag, we are faced with the problem that four different tracers of the gas indicate three different estimates of the total gas column density. The highest estimate is obtained from γ rays with the assumption of a "normal" C-R density. The middle one is obtained from the galaxy counts with ng^o=50, γ=0.75 and with the assumption of a normal gas-to-dust ratio. The lowest estimate is obtained from HI and CO with the assumptions of optical thinness for HI and $X=10^{20}$cm^{-2}K^{-1}km^{-1}s or from the star counts with the reference counts given by van Rhijn (1929) and the assumption of a "normal" gas-to-dust ratio. Clearly at least two of these assumptions must be invalid.

The difference between the two lowest estimates relies in part on the difference between the absorption measurements obtained with the use of star and galaxy counts and this is probably the first point to clarify. The absorption derived from the star counts is in fact the difference in absorption between the region of interest and a reference field.

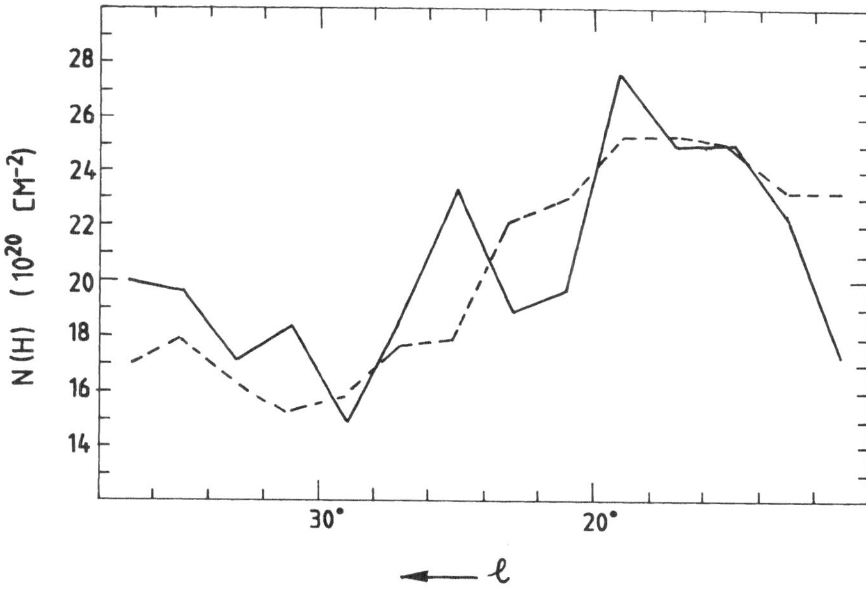

Figure 2. Oph-Sag. Longitude profile of the total gas column density averaged over the latitude range encompassed within the CO survey. The solid line is the estimation from the galaxy counts corrected for a possible overestimate of 0.25 mag (4 10^{20}cm^{-2}) and the dashed line represents the estimation based on CO (X = 10^{20}cm^{-2}K^{-1}km^{-1}s) and HI corrected by 25% for optical depth effects.

In the present case, the reference counts were taken from van Rhijn tables whose failings are well known (Racine 1969, Rossano 1978). It is almost impossible to find a field free of absorption in Oph-sag. Therefore it is quite likely that the star counts have underestimated the absorption. It is also possible that the absorption derived from the galaxy counts has been overestimated in Oph-Sag. As discussed by Heiles(1976), when the ratio of the galaxy to star number-density becomes too small, galaxies may be overlooked. This ratio in Oph-Sag is one of the lowest of the sky ($|b|>10°$) :10^3-10^4 .

Turning now our attention to the gas estimates from the radio measurements, we note that in the regions void of CO, the N(HI)/Apg(Ng) ratio is higher than the average in Oph-Sag, but still lower than the standard value. In addition to an overestimate of the absorption, there is three possibilities for such a difference:

i)the total gas to dust ratio is lower in Oph-Sag by nearly a factor of 2. However, the Copernicus measurements do not support this possibility. Excluding the ρ Oph star which shows an anomalously high N(H)/E(B-V) value, the average of 12 stars in the upper Scorpius and in Oph-Sag gives: N(H)/E(B-V) = 6.4 10^{21} cm^{-2} mag^{-1}, and for the 2 stars nearest to Oph-Sag: ζ Oph and 67 Oph, this ratio is unchanged, negligibly higher than the standard value.

ii) There is significant amounts of H_2 without any detected CO emission.

iii) The assumption of a negligible thickness for the 21 cm line does not hold in Oph-Sag, i.e. the average spin temperature is significantly lower than 120-130 K.

Although the second of these possibilities cannot be discarded, there is some indication from the 21 cm absorption measurements that indeed the third one is quite likely. There is 13 sources in Oph-Sag observed during the course of the Nancay survey of absorption towards extragalactic sources (Crovisier, Kazes and Aubry 1978), all of them show an absorption feature between 0 and 5 km s^{-1}, typical of the velocities found in the CO survey. The average absorption for these sources is 0.71 which correspond to an optical depth of 1.25 and imply a typical spin temperature of about 80 K. This indicates that N(HI) has been underestimated by about 25%. Since HI is the main component of this complex, consistency with the star counts estimate would be obtained by assuming a similar underestimate $\sim$ (0.5 mag) which is entirely reasonable. Now pushing further these estimates to reach those obtained from the galaxy counts seems unrealistic. This would require in particular HI absorption components much bigger than observed in the Nancay survey. An overestimate of the absorption derived from the galaxy counts of about 0.25 mag would achieve to reconcile all the radio and optical absorption estimates (see fig. 2) but around a level twice time lower than that indicated by the γ-ray data. Therefore, one has to consider that the γ-ray intensities observed in Oph-Sag do not result only from the interactions of a "normal" cosmic-ray flux with the interstellar gas. Three cases can be considered:

i) the C-R density in Oph-Sag is higher than elsewhere,

ii) Oph-Sag contains a population of weak γ-ray point sources,

iii) there is a galactic halo of inverse Compton emission.

This last possibility, as dicussed by Strong (1984) would imply also a symmetric γ-ray excess for negative latitudes. This is not observed. Riley and Wolfendale (1984) proposed an inverse Compton component proportional to the 408 MHz intensity which would reproduce the asymmetry of the γ-ray emission. However, this model encounters other difficulties, as discussed by Strong (1984). With the actual γ-ray data, it is not possible to test the second possibility. Such a study will await data from γ-ray experiments with better sensitivity and angular resolution and with a lower background level. The first possibility is very attractive and one cannot forget to notice the proximity of the radio continuum loop I whose expansion in the ISM could be accompanied by a shock-wave accelarating cosmic rays (Morfill et al. 1981). In this regard, Strong (1984) noted that if the North Polar Spur is mainly due to an enhancement of C-R electrons rather than to the compression of the magnetic field, as proposed by Heiles et al. (1980), the γ-ray enhancement could be entirely explained.

5 Acknowledgements

The author wish to thank T. Montmerle, A.W. Strong and J.A. Paul for many helpful comments on the present paper.

6 References

Baran, G.P. and Thaddeus, P. 1977, B.A.A.S., 9, 353.
Barnard, E.E. 1927, Photographic Atlas of Selected Regions of the Milky Way, ed. E.B. Frost, M.R. Calvert (Washington dc: Carnegie Institute of Washington).
Bloemen, J.B.G.M. 1984, Astr.Ap., in press.
Bloemen, J.B.G.M. et al. 1984, Astr.Ap., 139, 37.
Bignami, G.F., and Morfill, G.E. 1980, Astr.Ap., 87, 85.
Bohlin ,R.C., Savage, B.D., and Drake, J.F. 1978, Ap.J., 224, 132.
Burstein, D. and Heiles, C. 1978, Ap.J. 225, 40.
Caraveo, P. et al. 1980, Astr.Ap., 91, L3.
Caraveo, P. et al. 1981, Proc.17th Int.Cosmic Ray Conf., Paris 1, 139
Casse, M. and Paul, J.A. 1980, Ap.J., 237, 236.
Cernicharo, J. and Bachiller, R. 1984, Astr.Ap.Suppl., in press.
Crovisier, J., Kazes, I., and Aubry, D. 1978, Astr.Ap.Suppl., 32, 205.
Crovisier, J., Kazes, I., and Aubry, D. 1980, Astr.Ap.Suppl., 41, 229.
Dame, T.M. and Thaddeus P. 1984, in preparation.
Dickey, J.M., Salpeter, E.E., and Terzian, Y. 1978, Ap.J.Suppl., 36, 77.
Dickman, R.L. 1978, Ap.J.Suppl., 37, 407.
Encrenaz, P.J., Falgarone E. and Lucas R. 1975, Astr.Ap. 44, 73.
Fichtel, C.E., Simpson, G.A., and Thompson, D.J. 1978, Ap.J., 222, 833.
Garzoli, S.L., and Varsavsky, C.M. 1966, Ap.J., 145, 79.
Heiles, C. 1967, Ap.J.,
Heiles,C. and Habing H.J. 1974, Astr.Ap.Suppl., 14, 1.
Heiles, C. 1976, Ap.J., 204, 379.
Heiles, C. and Cleary, M.N. 1979, Aust.J.Phys.Suppl., 47, 1.
Heiles, C., Chu Y.-H., Reynolds, R.J., Yegingil I., Troland T.H. 1980, Ap.J., 242, 533.
Hubble, E.P. 1933, Mt.W.Contr., 21, 139.
Jenkins, E.B., and Savage B.D. 1974, Ap.J., 187, 243.
Knapp, G.R. and Kerr F.J. 1974, Astr.Ap., 35, 361.
Kutner, M.L. 1973, Molecules in the Galactic Environment, ed. M.A. Gordon and L.E Snyder (Wiley, New York)
Lebrun, F. and Paul, J.A. 1978, Astr.Ap., 65, 187.
Lebrun, F. and Paul, J.A. 1983, Ap.J., 266, 276.
Lebrun, F. and Huang, Y.-L. 1984, Ap.J., 281, 634.
Lebrun, F. et al. 1982, Astr.Ap., 107, 390.
Lebrun, F. et al. 1983, Ap.J., 274, 231.
Lilley, A.E. 1955, Ap.J., 121, 559.
Lucke, P.B. 1978, Astr.Ap. 64, 367.
Lundmark, K. 192: Upp.Obs.Medd., 12, 1.
Lynds, B.T. 1962, Ap.J.Suppl., 7, 1.
Maddalena, R.J., Morris, M., Moscowitz, J., and Thaddeus P., in preparation.
McCammon, D. et al. 1983, Ap.J., 269, 107.
McCuskey, S.W. 1938, Ap.J., 88, 209.
Meyers, P.C. 1975, Ap.J. 198, 331.
Morfill, G.E., Volk, H.J., Drury, L., Forman, M., Bignami, G.F. and Caraveo, P.A. 1981, Ap.J., 246, 810.
Neckel Th. and Klare G. 1980, Astr.Ap.Suppl., 42, 251.

Paresce, F. 1984, A.J., 89, 1022.
Penzias, A.A., Jefferts, K.B., and Wilson, R.W. 1971, Ap.J., 165, 229.
Perry C.L. and Johnston L. 1982, Ap.J., 50, 451.
Racine, R. 1969, A.J., 74, 1073.
Rossano, G.S. 1978, A.J., 83, 234.
Ryter C., Cesarsky, C.J. and Audouze, J. 1975, Ap.J., 198, 103.
Sanders, D.B., Solomon P.M. and Scoville N.Z. 1984, Ap.J. 276, 182.
Savage, B.D. and Jenkins E.B. 1972, Ap.J. 172, 491.
Shane, C.D. and Wirtanen C.A. 1967, Pub.Lick.Obs., 22, 1.
Strong, A.W. and Wolfendale, A.W. 1981, Phil.Trans.Roy.Soc.Lond.A. 301, 541.
Strong, A.W. and Lebrun, F. 1982, Astr.Ap., 105, 159.
Strong, A.W., Murray, J.D., Riley, P.A., and Osborne, J.L. 1982, M.N.R.A.S., 201, 495.
Strong, A.W. et al. 1982, Astr.Ap., 115, 404.
Strong, A.W. 1983, M.N.R.A.S. 202, 1025.
Strong, A.W. 1984, Astr.Ap., in press.
Sturch, C. 1969, A.J., 67,37.
Swanenburg B.N. et al. 1981, Ap.J. 243, L69.
Tucker, K.D., Dickman, R.L., Encrenaz, P.J. and Kutner, M.L. 1976, Ap.J., 210, 679.
van Hoof, A. 1969, B.A.N.Suppl., 3, 137.
van Rhijn, P.J. 1929, Publ.Kapteyn Astron.Lab., 43.
Weaver, H.,Williams, R.W. 1973, Astr.Ap.Suppl., 8, 1
Wesselius P.R. and Sancisi R. 1971, Astr.Ap., 11, 246.
Wilson, R.W., Jefferts, K.B., and Penzias, A.A. 1970, Ap.J., 161, L43.
Wouterloot, J. 1984, Astr.Ap., 135, 32.

MOLECULAR LINE OBSERVATIONS OF NEARBY DARK CLOUDS AT HIGH GALACTIC LATITUDES

K. Mattila, M. Toriseva and L. Malkamäki

Observatory and Astrophysics Laboratory, University of Helsinki
Tähtitorninmäki, 00130 Helsinki, Finland

WHY TO OBSERVE HIGH-LATITUDE DARK CLOUDS?

The purpose of this ongoing study is to investigate the connection between molecules, dust and radiation field inside dust clouds. Dark nebulae at high galactic latitudes appear most suitable for such an investigation for a number of reasons: 1) There is no confusion by fore- or background gas and dust clouds. 2) Since the high latitude nebulae are nearby objects (typically $r \lesssim 200$ pc), the ambient UV radiation field is obtained from the local OB star distribution (Jura, 1974). 3) Any associated UV or IR sources are more easily detected. 4) The incident radiation field is asymmetrical, especially in the UV where the main contribution comes from galactic plane O and B type stars. This asymmetry provides a potential means to better understand the influence of radiation on molecular formation and destruction processes. 5) It is possible to observe the light scattered by the dust in these nebulae against their dust-free surroundings and thus obtain additional information on the radiation field and on the properties of grains (Mattila, 1979; Mattila and Schnur, 1983).

A list of our cloud sample with coordinates, extinction and distance estimates is given in Table 1. We have carried out observations of these nebulae since 1975

Table 1. Basic data of our sample of dark nebulae at high galactic latitudes.

Nebula	l	b	A_B	r	Remarks
L134	4°.0	+35°.9	> 10^m	100±50 pc	
L1147/48 L1155/58	102.1	+15.4			
L1172	103.7	+13.9		350 pc	Reflection nebula NGC 7023
L1457	158.9	-34.6			
L1590	195.3	-16.6	5^m	300 pc	Reflection nebula v.d. Bergh 38
L1634	207.2	-23.5			
L1642	210.8	-36.7	3^m		Nebulous star + IRAS source
Cha I	297	-16	8^m	140 pc	Reflection nebulae Ced 110,111,112
L1778/1780	359.0	+36.8	4^m	100±50 pc	

in the spectral lines of HI (21 cm), OH (18 cm), CH (9 cm), H_2CO (6 cm) and CO (2.7 mm). In addition, optical observations have been made to determine the extinction, the surface brightness and in some cases also the distance of the clouds.

THE CHAMAELEON I DARK CLOUD - A NEARBY STAR FORMATION REGION IN THE SOUTHERN SKY

The Chamaeleon I dark cloud is a large cloud near the star ε Chamaeleontis (Hoffmeister 1962). It is one of the few nearby dark clouds suitable for detailed study of low mass star formation. Optical investigations (e.g. Aitken and Roche, 1981; Appenzeller, 1979; Glass, 1979; Grasdalen et al. 1975; Henize and Mendoza V, 1973) have revealed a large number of young and variable stars, 45 stars with Hα emission, and four Herbig-Haro objects (Schwartz, 1977; Schwartz and Dopita, 1980). Ground-based infrared observations have revealed highly reddened stars within the cloud (Hyland et al. 1982), and the IRAS satellite has found a number of embedded protostars (Baud et al. 1984). The Einstein observatory has detected several X-ray sources (Feigelson, 1984), believed to be pre-main-sequence stars with hot coronae.

Star counts have been carried out in the cloud using visual (Rydgren, 1980 and private communication) and blue Schmidt plates (Toriseva and Mattila, in preparation). The cloud mass obtained from our data is about 400 solar masses. The Schmidt plates show that the cloud is in a region of extensive thin dust, visible because of its extinction as well as reflection. It seems to be part of a dust layer parallel to the galactic disk and ~50 to 70 pc south of it (King et al. 1980).

We have made the first extensive radio line mapping of the Chamaeleon I dark cloud (Toriseva, Höglund and Mattila, 1984). The Parkes 64-m radio telescope was used to map the 6-cm formaldehyde absorption and the 18-cm OH emission. The map of the H_2CO peak antenna temperature distribution is shown in Fig. 1.

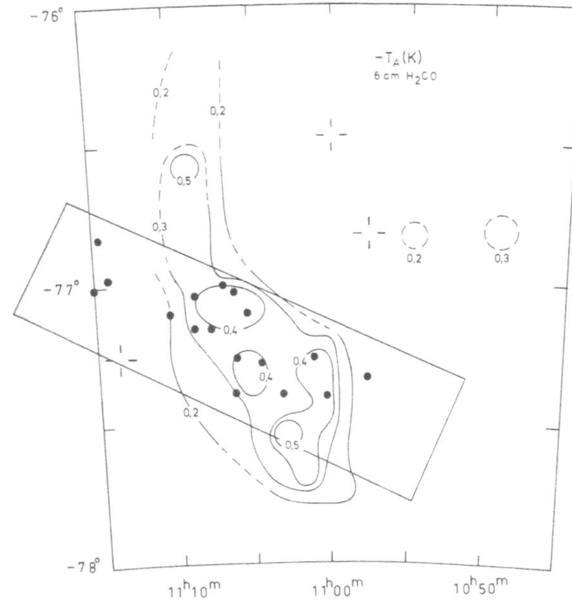

Fig. 1. The peak antenna temperature distribution of the 6 cm H_2CO line in the Cha I dark cloud. The far-IR sources found by the IRAS satellite, which fullfill the criterion F(60 μm) > F(25 μm), are shown as dots. The rectangle is the boundary of the high-sensitivity IRAS survey.

Comparison of the H_2CO and OH molecular line intensities with the extinction reveals a good correlation, and a similarity with the results for the Taurus cloud. There are three reflection nebulae associated with the cloud, two of them with B-type illuminating stars. There is no correlation (or anticorrelation) of the H_2CO line parameters with these nebulae. In Fig. 1 the IRAS sources with F(60 μm) > F(25 μm) are shown as dots. There appears to be some correlation with the formaldehyde peak positions (Toriseva and Mattila, in preparation).

CH OBSERVATIONS

CH is a widely distributed molecule in interstellar space and it is readily detected in dark nebulae with even a modest extinction of 1-2 magnitudes. We have carried out observations of the 3335 MHz (9 cm) CH line in L134, L1780 and L1590 using the 43-m radio telescope at Green Bank. CH column densities, N(CH), have been derived and are compared with (blue) optical extinction, A_B, derived from star counts and convolved with a Gaussian response function corresponding to the beam of the radio telescope. The observational results together with the least squares fits to the L1780 and L1590 data are shown in Fig. 2. A good correlation has been found between the CH column density and optical extinction, the correlation coefficient being $\rho = 0.83$ and 0.93 in L1590 and L1780, respectively. In both these clouds the relation is very similar, the mean value being:

$$N(CH) = 3.4 \times 10^{13} (A_B - 0^m\!.3) \ cm^{-2}$$

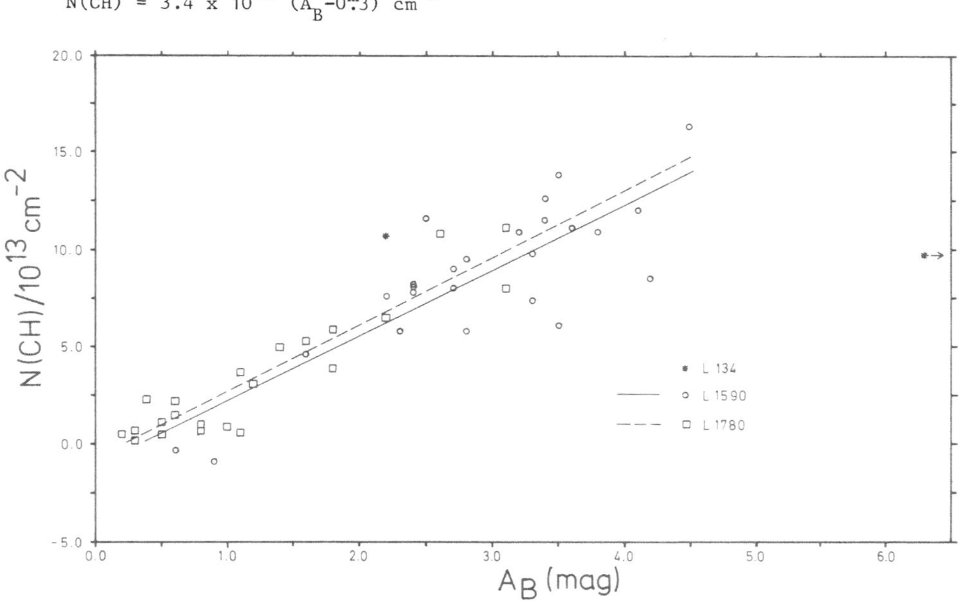

Fig. 2. Correlation between the CH column density and blue extinction with data for L134, L1590 and L1780 combined. The full line is the least squares fit for L1590 and the dashed line for L1780.

A comparison of this relationship with model calculations shows that the current CH formation theories with gas phase reactions only are able to explain the observations. We have also tried to detect differences of the CH abundances between the bright and shadow sides of the clouds L1780 and L1590, but with negative results. For a more detailed report see Mattila (1984).

NEUTRAL HYDROGEN IN THE PERIPHERIES OF DARK CLOUDS

The theory of H_2 formation on grain surfaces and destruction by UV photons predicts that inside dust clouds, where $A_V \geqq 1^m$, almost all hydrogen has been converted into H_2 (see e.g. Hollenbach et al. 1971). Atomic hydrogen is predicted to be confined to a narrow layer with $A_V < 1^m$ at the surface of the cloud. High-latitude dark clouds are specially suited to look for this HI gas since the background fluctuations of the 21-cm emission are here much less severe than at low latitudes. Furthermore, the atomic gas is expected to be present predominantly on the bright side of a high latitude nebula (facing the galactic plane) where the dissociation of H_2 by UV photons is strongest. We first detected this excess HI emission in the dark cloud L1780 (Mattila and Sandell, 1979). Wannier et al. (1983) have observed similar warm HI halos around a few other molecular and dark clouds.

We have observed, in addition to L1780, also the clouds L1147/1148/1155/1158, L1172, L1457, L1590, L1634 and L1642 in the HI 21-cm line using the Effelsberg 100-m telescope (HPBW = 9'). So far the data for L1590 and L1642 have been reduced and indicate the presence of excess HI emission in the periphery of the dust distribution of these clouds. This excess emission is in both these cases located near the 0.5-1 magnitude extinction and on the bright side of the cloud, facing the galactic plane.

CO OBSERVATIONS

We have mapped L1155/58, L1457, L1590, L1634, L1642 and L1780 in the ^{12}CO (J = 1) line using the GISS 1-m radio telescope in New York. Partial maps of the central areas and observations of selected positions also in the isotopes ^{13}CO and $C^{18}O$ have been made using the Kitt Peak 11-m telescope. Furthermore, the 14-m telescope of the FCRAO was used to map the location of the IRAS-source in L1642 (IRAS Circular No. 1, Astron. Astrophys. 123, C1, 1983) both in the ^{12}CO (1-0) and ^{13}CO (1-0) line. Neither a CO hot spot nor a density peak was detected (Mattila and Friberg, private communication).

REFERENCES

Aitken,D.K., Roche,P.F.: 1981, Monthly Notices Roy. Astr. Soc. 196, 39 p

Appenzeller,I.: 1979, Astron. Astrophys. 71, 305

Baud,B., Young,E., Beichman,C.A., Beintema,D.A., Emerson,J.P., Habing,H.J., Harris,S., Jennings,R.E., Marsden,P.L., Wesselius,P.R.: 1984, Astrophys. J. 278, L53

Feigelson,E.D.: 1984, X-ray emission from pre-main-sequence stars, in Lecture Notes in Physics, Vol. 193, Springer-Verlag, p. 27

Glass,I.S.: 1979, Monthly Notices Roy. Astr. Soc. 187, 305

Grasdalen,G., Joyce,R., Knacke,R.F., Strom,S.E., Strom,K.M.: 1975, Astron. J. 80, 117

Henize,K.G., Mendoza V.,E.E.: 1973, Astrophys. J. 180, 115

Hoffmeister,C.: 1962, Zeitschrift für Astrophys. 55, 290

Hollenbach,D., Werner,M.W., Salpeter,E.E.: 1971, Astrophys. J. 163, 165

Hyland,A.R., Jones,T.J., Mitchell,R.M.: 1982, Monthly Notices Roy. Astr. Soc. 201, 1095
Jura,M.: 1974, Astrophys. J. 191, 375
King,D.J., Taylor,K.N.R., Tritton,K.P.: 1980, Monthly Notices Roy. Astr. Soc. 188, 719
Mattila,K.: 1979, Astron. Astrophys. 78, 253
Mattila,K.: 1984, submitted to Astron. Astrophys.
Mattila,K., Schnur,G.: 1983, Mitteilungen der Astron. Gesellschaft 60, 387
Mattila,K., Sandell,G.: 1979, Astron. Astrophys. 78, 264
Rydgren,A.E.: 1980, Astron. J. 85, 444
Schwartz,R.D.: 1977, Astrophys. J. Suppl. 35, 161
Schwartz,R.D., Dopita,M.A.: 1980, Astrophys. J. 236, 543
Toriseva,M., Höglund,B., Mattila,K.: 1984, Proc. III Latin-American Reg. Meeting in Astronomy, Buenos Aires (in press)
Wannier,P.G., Lichten,S.M., Morris,M.: 1983, Astrophys. J. 268, 727

Comparison of Optical Appearance and Infrared Emission of Extended Dust Clouds.

C.P.de Vries
Sterrewacht Leiden
Netherlands

1.Abstract. Several methods exist to observe interstellar dust. In this paper, some regions are selected of which optical appearance, like extinction and surface brightness, are discussed, together with a prelimenary comparison of infrared (100 μm) emission. Careful comparison of extinction, reflection and IR-emission can yield information on the radiative properties of interstellar dust and of the interstellar radiation field.

2.Introduction.

The IRAS-Infrared survey has provided us with a complete map of Infrared emission. Since dust transmits it's energy in the infrared, this survey will enable us to study the whole sky distribution, temperature and density of cosmic dust. However, a good understanding of the physical properties of dust is necessary. Therefore a comparison with other (means of) observations of dust is of extreme interest.

Dark matter can be mapped by means of starcounts. Also there is a relation between the optical depth of a dark cloud and its surface brightness (see e.g. Mattila 1979). This enables us to detect thin dust clouds at high galactic latitudes with a higher efficiency and resolution then would be possible from starcounts only. Looking at thin clouds in simple radiation fields will make analysis of dust properties easier, since temperatures and radiation fields will be constant throughout the cloud.

In the next section, surface brightness and extinction in a few selected areas are compared. For this purpose, photografic plates were scanned using the Leiden Observatory "Astroscan" measuring machine. The data were reduced using an especially developed starcounting program which produces maps of both stellar density and surface brightness. This program is decribed elsewhere (de Vries (1984)). Extinction was derived, using the van Rhijn tables (1929).

3.Maps of a few selected area's.

a) Taurus region. Fig 1a shows a surface brightness map (blue PSS-plate) of a region in the Taurus dark clouds at appr. $l=175^{o}$, $b=-19^{o}$ At places (bottom,middle left) of large surface brightness, a few reflection nebulae can be recognized. Comparing absolute to apparent magnitudes of the stars, associated with the nebulae, yields two sets of distances of 40 and 180 pc.

A map of extinction is shown in fig 1b (black means high extinction). Combination with the previous map (fig 1c), shows that a relation exists between surface brightness and extinction. This relation is plotted in fig 2. The discrete nature of starcounting causes the apparent discretisation in the extinction-plots.

The wings of the brighter stellar images cause high surface brightnesses at average extinction (the peak arround 0.6 magnitudes extinction). Ignoring this, a relation between surface brightness and extinction can be drawn (the line in fig 2). At low extinctions, the surface brightness increases with extinction, up to a maximum arround 1.6-1.8 magnitudes of extiction, where the surface brightness starts to decrease again.

b) High latitude thin dust. Fig 3 shows a thin cloud at $l=314^o$, $b=-25^o$ which was found by searching the (IIIa-J) ESO/SERC sky-survey for excess surface brightness.

The relation found in fig 2 also holds for this region (fig 4), although the general interstellar radiation field probably differs. The fact that the surface brightness of this cloud mainly increases with extinction confirms the optically thin nature of this dust. The detailed structure of this cloud can therefore be seen best, by using the surface brightness.

c) λ-Orionis region. 5a shows the region arround λ-Orionis, at $l=195^o$,$b=-12^o$ on the IRAS 100μm sky-survey map. The position of λ-Orionis is indicated with a cross. This area is estimated at a distance of 400 pc. (P. Murdin and M.V. Penston (1977)). A reflection nebula of warm dust close to λ-Orionis shows clearly as an increase in 100-micron flux. Fig 5b shows the extinction in this area and fig 5c combines the infrared and optical data.

As can be expected, an excellent agreement is found, confirming the relation between extinction and the infrared flux of cosmic dust.

4. Conclusions. Extinction and surface brightness of dust clouds are related, showing a maximum surface brightness for (high latitude clouds) at approximately 1.6 magnitudes extinction. This fact can be used in finding and studying thin clouds at high galactic latitudes. Due to their expected (locally) simple radiation fields and optically thin character these clouds seem most suited for studies of the radiative properties of dust.

Acknowledgements: The author wants to thank the staff of UKSTU for kindly making available the photographic plate of the cloud-area in fig 3.

References

Mattila, K., 1979, Astron. & Astroph. 78, 253
Murdin, P., Penston, M.V., 1977, M.N.R.A.S. 181, 657
van Rhijn, P.J., 1929, Publ. Kapteyn Astron. Lab., No. 43
de Vries, C.P., 1984, Astronomical Photography 1984, Edinburgh ed. E. Sim and K. Ishida p. 311

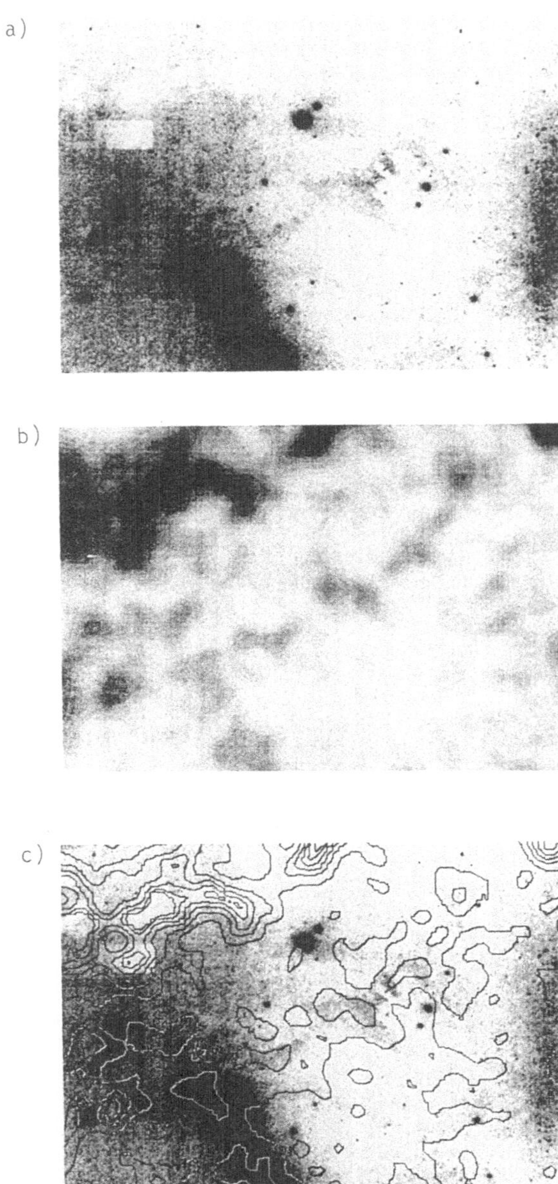

Fig.1. A region in the Taurus dark clouds around position l = 175°, b = -19°
(a) Surface brightness. (The small, bright square area upper left is due to an error in the scanning procedure.)
(b) Extinction as derived from starcounts.
(c) Contours of extinction overlay on surface brightness. Contour increment is 0.5 magnitude.

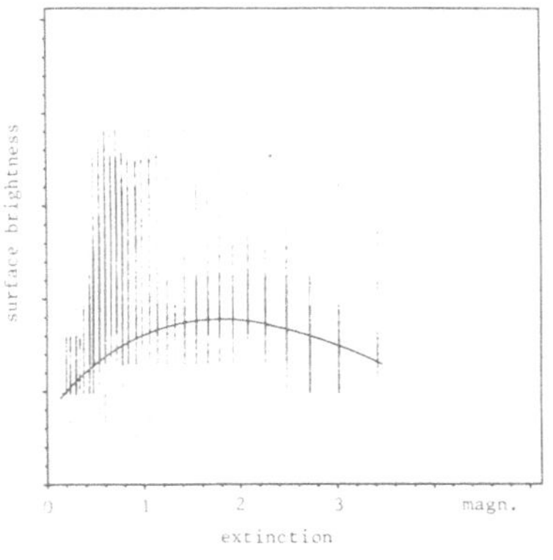

Fig.2. Relation between surface brightness and extinction in the Taurus region

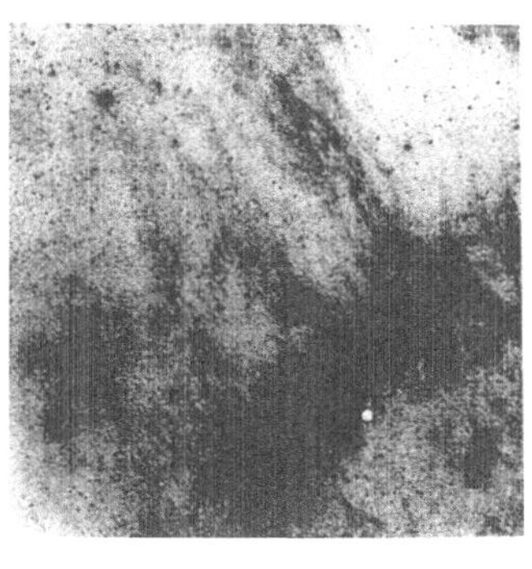

Fig.3. Thin dust cloud at l = 314°, b = -25°. Surface brightness map

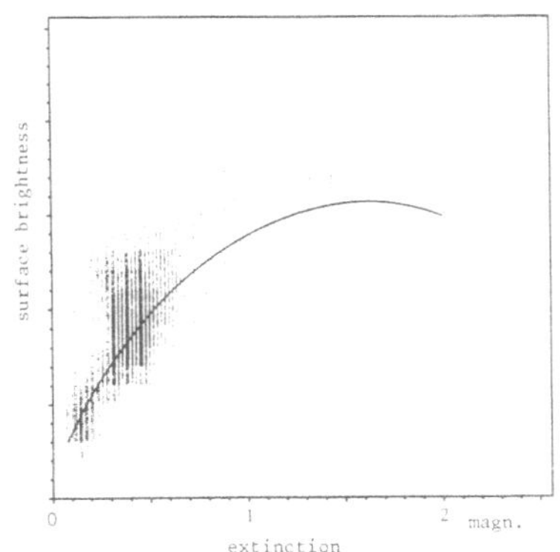

Fig.4. Relation between surface brightness and extinction of the cloud in Fig.3

a)

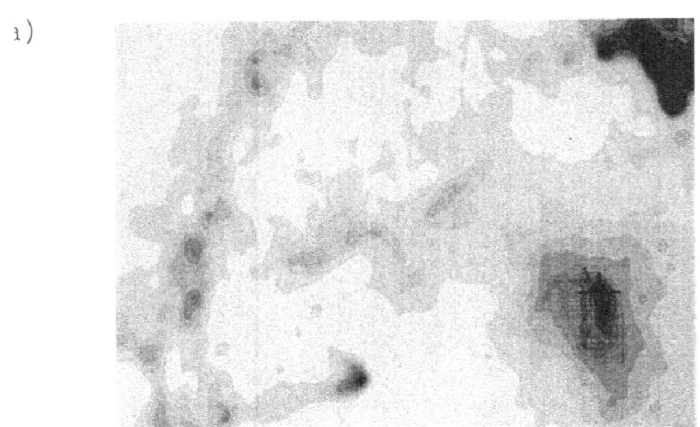

b)

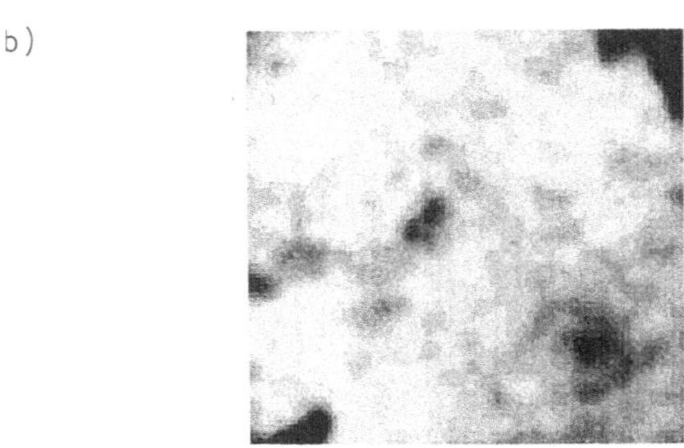

c)

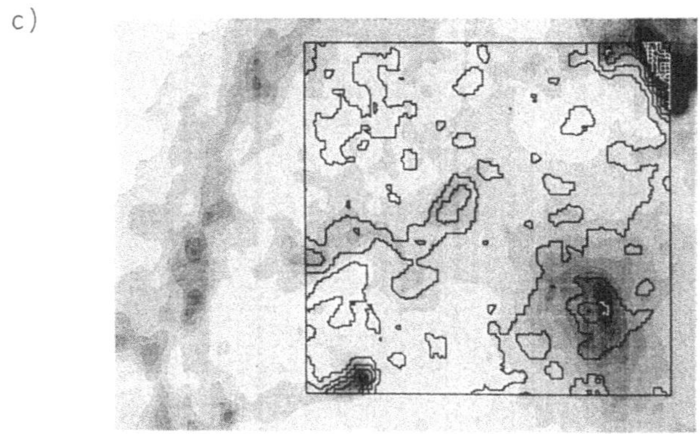

Fig.5. The region around λ-Orionis.
(a) 100 micron IRAS map
(b) Extinction as derived from starcounts
(c) Contours of extinction overlay on the infrared map. Contour increment is 0.3 magnitude

CORRELATIONS BETWEEN NH_3 (1.1) AND FAR IR EMISSION

L.G. Stenholm,[1] B. Baud,[2] R. Mauersberger,[3] K.M. Menten[3]

[1]Stockholm Observatory
S-133 00 Saltsjöbaden
Sweden

[2]Kapteyn Laboratory
Postbus 800
9700 AV Groningen
The Netherlands

[3]Max-Planck-Institut für Radioastronomie
Auf dem Hügel 69
D-5300 Bonn 1
Fed. Rep. of Germany

Introduction

The successful IRAS project (see Neugebauer et al, 1984) has given us a large quantity of IR data on premainsequence objects deeply imbedded in molecular clouds. This paper is a pilot project on the interaction between these newly formed stellar objects and the surrounding gas.

Present theories of star formation predict that protostars form out of the denser parts of molecular clouds. The first step is the formation of a gravitationally bound fragment. It will slowly contract sheding angular momentum through production of Alfvén waves (Norman 1984). Later on, the field decouples from the fragment and a stellar object forms. The protostar will deposit radiative and mechanical energy into the gas.

The evolutionary stage of a forming star can thus be deduced from its way of interacting with the gas. We intend to search for signs of such interactions and we will briefly discuss their physical significance.

Observations

We selected six IRAS sources characterized by $S_{100} > S_{60}$ (i.e. cool) and by a spatial coincidence with strong obscuration. The sources are 0412 + 180 (IRAS cir 1,8 and Emerson et al 1984) and all are in the vicinity of the Taurus region.

The positional accuracy for these sources is better than 45" x 20".

NH_3 data were aquired at the 100-m Effelsberg telescope with the 1024 channel autocorrelator and the K-band maser amplifier. We mapped the six regions simultaneusly in the NH_3 (1,1) and (2,2) lines. Each map is a 9 point square map centered on the IRAS source position. A spacing of 40" was used to ensure that the IRAS source is included in the map. Spatial resolution (HPBW) for NH_3 is 40" and the velocity resolution used was 0.16 km s^{-1}. Typical intensity errors in the spectra are 0.1 K.

Results and Discussion

NH_3 emission is found in the direction of all six sources. The intensity range is $0.15\,K \leq \bar{T}_A^* \leq 4.1\,K$. $\bar{T}_A^*$ is the mean of the line peak intensity from all 9 map positions.

The fact that a low luminosity protostar locally generates extra heat and gas motions has seldomly been proven by observations (except in the form of bipolar outflows). This is understandable since molecular clouds are compressive and turbulent (Stenholm 1984). A large number of eddies will thus exist along a line of sight. The emission from a young object is thus strongly contaminated by radiation from normal gas elements.

A relation between T_A^* (peak antenna temperature) and ΔV linewidth can be derived for a compressible medium. We thus derive that

$$\sqrt{T_A^*}\,\Delta V^{5/2} = \text{const.} \tag{1}$$

The specific relation given, rests on the assumption that the mean physical conditions over a map are similar. Further we assume that the kinetic energy in a volume element is conserved. Lastly $\tau < 1$ and subthermal extitation is assumed. The last conditions are less important since they do not change the basic T_A^*, ΔV anticorrelation of the compressible gas.

The maps of the two strong NH_3 sources, B 5 (0344+327) and L1551 (0422 + 180) , (which have good signal to noise ratios and very accurate IRAS positions within a few arcseconds) verifie the suggested anticorrelation between T_A^* and ΔV. In addition we see a peak $\sqrt{T_A^*}\,\Delta V^{5/2}$ value for the NH_3 emission at the position which coincides with the IRAS source. The gas near the protostar is thus energized by the young object. The 0412 + 261 source shows a similar behaviour. The signal to noise ratio of the other three maps is to low to give any conclusive evidences for either compression or protostellar-gas interaction.

The interaction between the young objects and the clouds is also seen as a correlation between the NH_3 intensity and the 100 μ flux density (S_{100}).

$$\log S_{100} = 1.64 + 1.18 \log \bar{T}_A^* \tag{2}$$

Correlation coefficient is 0.85.

This relation can be explained as a heating of the gas by dust around the protostar. A relation between dust colour and $\bar{T}_A^*$ is thus expected. The relation were found to be

$$\log \frac{S_{60}}{S_{100}} = -0.355 + 0.763 \log \bar{T}_A^* \tag{3}$$

the correlation coefficient is 0.91.

The influence of the protostar on the gas dynamics is less apparent in the mean linewidth of a map ($\overline{\Delta V}$) than in the intensity. There is thus no S_{100}, $\overline{\Delta V}$ correlation We found, however, a relation between the 100 μ flux density and the linewidth fluctuations.

$$\log S_{100} = -0.546 - 2.86 \log (\sigma_{\Delta V}/\overline{\Delta V}) \tag{4}$$

where $\overline{\Delta V}$ is the mean linewidth (FWHM) of a map and $\sigma_{\Delta V}$ the standard deviation of the linewidths within a map. The relation has a correlation factor of -0.84.

The physical cause of this relation is unclear. We suggest that it shows that an efficient starformation occurs in those cloud regions, where the gasmotions have been suppressed by efficient magnetic angular momentum transport.

The absence of a relation for $\overline{\Delta V}$ itself (which represents the small scale end of gas motions) is surprising. One hint is given by Myers 1984 who finds a lower $\overline{\Delta V}$ in cloud cores without IR sources than in those with sources. Our suggestion is that all scales of gas motions were depressed just before the formation of the IR sources. The gas has then subsequently aquired some kinetic energy from the protostars which preferentially increased the small scale motions i.e. ΔV itself. Two conflicting processes have thus been at work at the small scale end of the gas motions. The larger scales on the other hand have only been reshaped by the angular momentum transport mechanism.

Acknowledgement

LGS acknowledges the support of the MPI for Radio Astronomy Bonn and of the project "Theoretical studies of star formation, F-FU 3983-101" which is sponsored by the Swedish Natural Science Research Council.

References

Emerson, J.P., Harris, S., Jennings, R.E., Beichman, C.A., Baud, B., Beintema, D.A., Marsden, P.L., Wesselius, P.R., 1984, Astrophys. J. Lett. 278, L49

Myers, P.C., 1984, Europhysics conf. abst. 81, 56

Neugebauer, G., Habing, H.J., van Duinen, R., Aumann, H.H., Baud, B., Beichman, C.A., Beintema, D.A., Boggess, N., Clegg, P.E., de Jong, T., Emerson, J.P., Gartier, T.N., Gillett, F.C., Harris, S., Hauser, M.G., Houch, J.R., Jennings, R.E., Low, F.J., Marsden, P.L., Miley, G., Olnon, F.M., Pottasch, S.R., Raimond, E., Rowan-Robinson, M., Soifer, B.T., Walker, R.G., Wesselius, P.R., Young, E., 1984, Astrophys. J. Lett. 278, L1

Norman, C., 1984, Mem.Soc.Ast. Italiana to be published

Stenholm, L.G. 1984, Astron. Astrophys. 137, 133.

Correlation of high galactic latitude HI and FIR emission and warm dust in the neutral interstellar medium

F. Boulanger, B. Baud, G.D. van Albada
Groningen Space Research Department
Postbus 800, 9700 AV Groningen, The Netherlands.

Abstract: A comparative study of HI and IRAS maps of a large field at high galactic latitude is presented. The galactic infra-red emission is shown to be well correlated with the HI distribution. A quantitative relation between HI column densities and long-wavelengths IRAS fluxes is derived.

The infra-red spectrum of one diffuse HI cloud, from 12 to 100 μm is presented. Its emission in the mid infra-red (12 and 25 μm) is several order of magnitudes larger than what is predicted by standard dust models. The power radiated at 12 μm represents 25% of the 100 μm emission. The spectrum presents a minimum at 25 μm. The color temperature derived from the 12 and 25 μm fluxes is 320K.

1. Observations and data reduction

We used for this investigation IRAS survey observations and HI data from Heiles and Habing (1974). The correlation between HI and infra-red emission presented in section 2 is based on 60μm and 100μm IRAS observations of a $20^{\circ}\times18^{\circ}$ (α,δ) field which were processed at the Groningen Space Research Department. In order to derive the spectrum of the high galactic latitude extended sources (Cirrus) we used a set of $18^{\circ}\times16^{\circ}$, 12, 25, 60 and 100μm maps (Sky Flux Images) of a neighbour field, provided by the Jet Propulsion Laboratories(Gautier et al. 1984). All IRAS maps were smoothed to the 36' resolution of the HI observations. Before smoothing, all bright point sources were subtracted from the original maps.

In order to obtain from IRAS observations the galactic component of the emission, it is necessary to model and subtract the zodiacal emission (ZE). The detailed procedure of this subtraction is described in Boulanger et al. (1985).

All parameters of the computed ZE fits are summarized in table 1. The 60 and 100μm galactic emission maps used in section 2 are displayed in figure 1-a and b. The HI data integrated from -45 km/s to +45 km/s were interpolated on a grid corresponding to the infra-red maps. The resulting map is displayed in figure 1-c. Infra-red and HI maps do not overlap completely since there is a 1.2° gap in the HI observations at a galactic latitude of 72°. In the cross scan direction, the structure in the Sky Flux Images is distorted by stripes of peak $\lambda \times I_{\lambda}$ amplitude of few 10**-8 W/M2/Sr. These maps will therefore only be used to obtain cuts along the scan direction. The 100µm Sky Flux Image that will be discussed in section 3 is displayed in figure 2.

2-Correlation between far infra-red and HI emission

The maps displayed in figure 1-a, b and c show a clear correlation between the HI distribution and the infra-red emission. All main features of the infra-red maps have a counterpart in the HI distribution. The 100µm brightness and HI column densities relative to the respective background emissions were measured for all cirrus marked by a cross in figures 1-a, b and c. All measurements are consistent with an 100 µm emission to HI column density ratio:

I_{λ}(100µm) (Jy/Sr)/ NHI(H/cm2) = $1.4+.3 10^{-14}$

Assuming for the HI clouds in the field, a normal A_{V} /N_{H} ratio of $5.3 10^{-22}$ mag cm2 (Bohlin et al. 1978, Savage and Mathis 1979), we derive the following relation between A_{V} and the 100µm emission: A_{V} = .038+.008mag*I_{λ}(100µm)(MJy/Sr)

Hauser et al. (1984) estimated the smooth galactic component at the south galactic pole flux at 100µm to be 2.4 + 1 MJy/Sr. If we assume that the relation we derived for the cirrus component applies to the smooth galactic background, we derive an extinction at the galactic poles A_{V} = .09+.04.

The correlation observed between the HI column densities and the infra-red fluxes implies that dust abundance and properties are rather uniform over the figure 1 field. This correlation suggests also that it is possible to use the IRAS observations to map the dust distribution. At 36' resolution it is possible to map λI_{λ}(100µm) variations as small as $5*10^{-9}$ W/M2/Sr. According to the relation derived above, this brightness corresponds to an HI column density of $1.5 * 10^{19}$ H/cm2 and an visual extinction Av= 0.01 for normal dust abundance and extinction properties.

3-The spectrum of the infra-red emission of diffuse HI clouds

The cirrus marked by a cross in the 100 µm map displayed in figure 2 corresponds in the HI survey to a well defined cloud of HI column density $3.3*10^{20}$ H/cm2. This HI column density corresponds to A_{V}=0.17 for the average solar dust abundance and extinction curve.

In figure 3, the HI, 12, 25, 60 and 100µm emission profiles across this cloud along the cut parallel to the scan direction shown in figure 2 are displayed. This HI cloud is clearly detected at all four wavelengths. The corresponding spectrum is plotted in figure 4.

Its emission at 60 and 100µm is consistent with the emission of the other HI

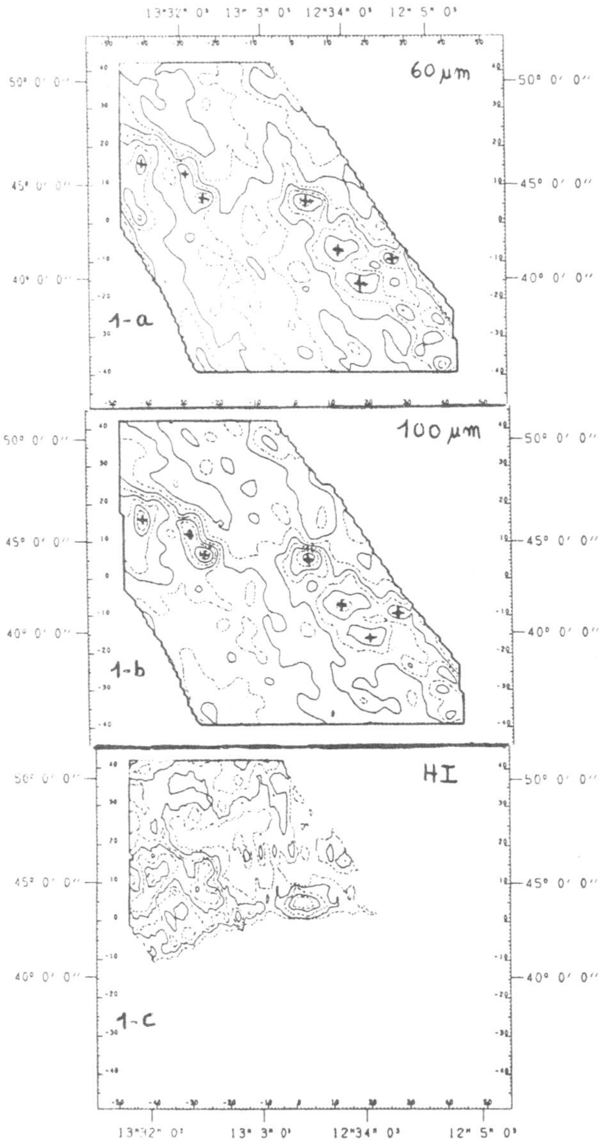

Fig.1a-c. 60, 100 μm and HI maps of a 20° x 18° field. The contours in the infra-red maps represent fluxes integrated over the IRAS bands. The spacing between contours is $2 \cdot 10^{-9}$ $W/m^2/sr$ for the 60 μm map and $2.5 \cdot 10^{-9}$ for the 100 μm map. The HI contours go from 50 K km/s to 110 K km/s by steps of 10 K km/s

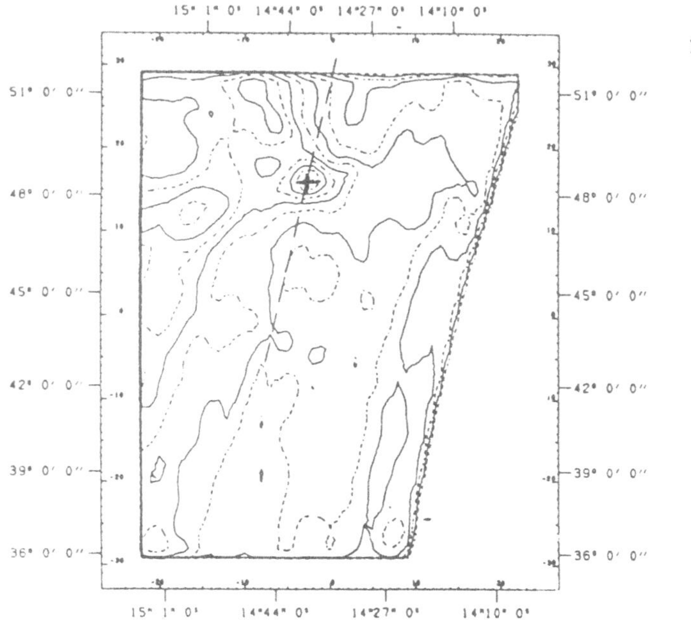

Fig.2. 100 µm Sky Flux Image. The contours represent the flux integrated over the 100 µm band. They are linearly spaced by steps of $5 \cdot 10^{-9}$ $W/m^2/sr$

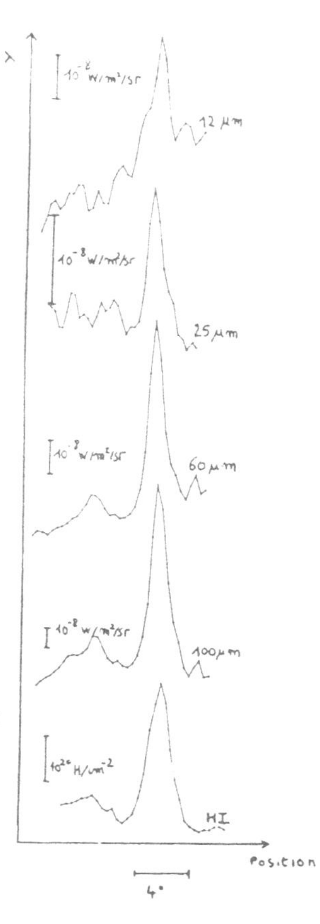

Fig.3. Cuts across the cirrus centered at α = 14h40m and δ = 48° in the HI and infra-red maps displayed in Fig.1. The direction of the cut is displayed in Fig.2

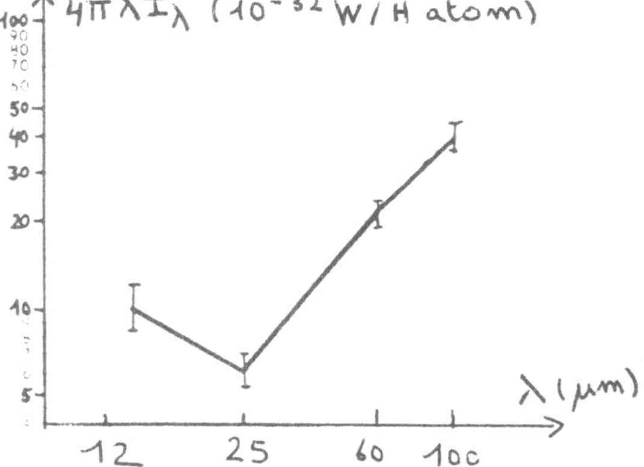

Fig.4. Infra-red emission spectrum of the cirrus marked by a cross in Fig.2. The spectrum has been normalized per H atom

clouds studied in the previous section. Therefore its overall spectrum from 12 to 100μm is likely characteristic of the emission of interstellar dust heated by the local interstellar radiation field. The emission observed at 12 and 25μm is several order of magnitudes larger than what one expects from emission of dust grains heated by the interstellar radiation field radiating at their equilibrium temperature (Mathis et al. 1983, Draine and Hyung Mok Lee 1984). The color temperature computed for a constant emissivity increases gradually from the long to the short wavelengths:

T(60μm,100μm)= 32K , T(25μm,60μm)= 75K, T(12μm,25μm)= 320K

The most obvious explanation of the excess emission at 12 and 25μm, is the existence of extremely small grains which are transiently heated to high temperatures each time they absorb one photon (Purcell 1976, Selgrenn 1984, Puget et al. 1984). Leger and Puget (1984) have suggested that the existence of large carbon molecules could explain the unidentified infra-red features. The existence of small graphite grains with sizes approaching those of these large molecules can also account for the two main features of the spectrum that we observe: the large ratio of 10 to 100 μm emission and the minimum at 25 μm due to the presence of their emission features in the 12μm IRAS band (Puget et al. 1984).

Table 1

Zodiacal emission fits: $\lambda \times I_\lambda$ = a(W/M2/Sr)/Sin(β)

	λ	a		λ	a
Figure 1, field:	60μm	3.7e-7	Sky Flux images:	12μm	4.9e-6
	100μm	0.8e-7		25μm	4.1e-6
				60μm	5.2e-7
				100μm	1.1e-7

References:

Bohlin R.C., Savage B.D., Drake J.F., 1978, Astrophys. J. 224,132

Boulanger F., Baud B., van Albada G.D., 1985, In preparation.

Gautier T.N. et al., 1984, in preparation

Hauser M. et al., 1984, Astrophys. J. 278,L15

Heiles C., Habing H.J. 1974, Astron. Astrophys. Suppl. 14,1

Leger A., Puget J.L. 1984, Astron. Astrophys. 137,L5

Low F.J. et al., 1984, Astrophys. J. 278,L19

Mathis J.S., Mezger P.G., Panagia N., 1983, Astron. Astrophys. 128,212

Puget J.L., Leger A., Boulanger F., 1984, Astron. Astrophys. to be published

Purcell E.M., 1976, Astrophys. J. 206,685

Savage B.D., Mathis J.S., 1979, Ann. Rev. Astron. Astrophys. 17,73

Selgrenn K., 1984, Astrophys. J. to be published

Galactic molecular clouds: their size and mass distribution

H. Zinnecker
Royal Observatory, Edinburgh

and

S. Drapatz
Max-Planck-Institut für extraterrestrische Physik, Garching

Abstract

By combining observational results on dwarf and giant molecular clouds, we deduce the following expressions for their size and mass distribution:

(1) $\log(dn/dr) = 4.3 - 0.65 \log^2(r/0.1\ \mathrm{pc}) \quad r \geqslant 0.1\ \mathrm{pc}$
(2) $\log(dn/d\log M) = 4.0 - 0.16 \log^2(M/132\ M_o) \quad M \geqslant 2\ M_o$

These expressions are suggested to hold in the spiral arms of the Galaxy within galacto-centric distances 4.5 Kpc $<$ R $\leqslant$ 10 Kpc. In the interarm regions they must be truncated at $r \leqslant 15$ pc and $M \leqslant 4\times10^4\ M_o$.

I. Introduction

Do giant molecular clouds (GMCs, Solomon & Sanders 1980) form from dwarf molecular clouds (DMCs, Elmegreen 1984 a,b) or vice versa? In other words: which clouds come first, the small ones or the large ones? Thus, in terms of models, the question is which type of models for cloud formation is likely to be correct - coagulation models (Field & Saslaw 1965, Kwan 1979, Hausman 1982, Kwan & Valdes 1983) or fragmentation models (Parker 1966, Mouschovias, Shu & Woodward 1974, Elmegreen 1979, Blitz & Shu 1980)? Of course, hybrid models involving growth and disruption at the same time are also conceivable (cf. Chièze & Lazareff 1980, Casoli & Combes 1982).

If observations can tell the size and mass distribution of molecular clouds over a sufficiently wide range in size and mass, it may be possible to infer which mode of molecular cloud formation is operational. Therefore, it seems promising to deduce molecular cloud spectra from observations of a complete sample covering both GMCs as well as DMCs (i.e. Lynds dark clouds). An attempt towards this end has been made in a recent paper of ours (Drapatz & Zinnecker 1984, hereafter DZ84).

II. Fitting the observations

We shall not repeat the whole DZ84 paper. It suffices to reproduce Table 1 which gives the parameters for the distributions dn/dr of cloud radii r in different size-ranges. Here we have assumed that the different observations can be fitted approximately by power laws of the form

$$\log(dn/dr) = A - B \log r \qquad (1)$$

(dn/dr is the number of molecular clouds per Kpc^3 per unit radius in pc). The reason why we believe we can combine the data sets of DMCs and GMCs to construct a complete sample lies in the fact that the limiting sensitivity or column density for DMCs agrees with that for GMCs (see DZ84). The cloud-size distributions of the different observations are usually provided for certain areas of the galactic plane. They have been normalized by dividing them by the ratio of mean H_2 densities obtained for these areas to the corresponding densities of the area between galacto-centric distance R = 4.5 and 10 Kpc.

Table I. Parameters for the distribution of cloud-sizes in different size-ranges

A	B	r_{min}	r_{max}	Ref.
3.6	1.0	0.1	10	Rowan-Robinson (1979)
3.1	1.8	2.5	37	Stark (1979)
3.8	2.2	5.0	55	Solomon & Sanders (1980)
5.2	3.3	9.0	30	Liszt et al. (1981)

Realizing that the power B in Table 1 increases systematically with r_{min}, the resulting fit to the size distribution is

$$\log(dn/dr) = 4.3 - 0.65\,(1+\log r)^2 \qquad r \geq 0.1 \text{ pc}, \qquad (2)$$

resembling a half log-normal distribution.

Moreover, observations suggest the following molecular cloud mass/radius relation

$$M(M_\odot) \approx 200\ r^2(\text{pc}), \qquad (3)$$

since the cloud radii are found to scale inversely with the particle density in the molecular clouds (Cowsik, Drapatz & Michel 1975, Rowan-Robinson 1979, Larson 1981). Making use of this relation to transform the cloud size spectra into cloud mass spectra yields

$$\log(dn/d\log M) = 4.0 - 0.16(-2.12+\log M)^2 \qquad M \geqslant 2\ M_o \quad (4)$$

Note that the rms deviation in log M is roughly 2.6 times the corresponding rms deviation of the log-normal stellar IMF (Miller & Scalo 1979, Zinnecker 1984).

The cloud size or mass distribution can also be taken to derive the distribution of the cloud velocity dispersion (We are indebted to Dr. J. Canto for this suggestion.) Using Larson's empirical (1981) result

$$\sigma(\mathrm{km/s}) = 0.42\ (M/M_o)^{0.2} \qquad (5)$$

where σ is the three-dimensional velocity dispersion, we obtain

$$\log \frac{dn}{d\sigma} = 4.3 - 4\ \log^2(\ \sigma/0.84\ \mathrm{km\ s^{-1}}) \qquad (6)$$

Finally, the cloud size distribution allows to calculate the number of shocks due to cloud cloud collisions in the Galaxy which may be observed by their infrared emission (Drapatz 1983).

III. Conclusion and future tests

Figure 1 displays the distributions that we determined. We suggest that these distributions are valid throughout the range 4.5 Kpc < R < 10 Kpc in galacto-centric distance with the limitation that the difference between the arm and interarm regions is entirely due to the truncation of the distribution for $r > 15$ pc (hence $M > 4.5\times10^4$ M_o according to equ.(3)) in the interarm region (Stark 1983). A model in which the primary cloud units are the most massive clouds, with smaller sized clouds resulting from the successive disruption of larger clouds by OB star formation or other processes, might be consistent with the observed cloud-size and -mass distribution.

An interesting test of the size and mass distributions derived here would be to study in detail the clumpiness and scales inside a single giant molecular cloud (cf. Felli et al. 1984), in a spirit similar to studies of the IMF of open galactic clusters complementing the determination of the field star IMF.

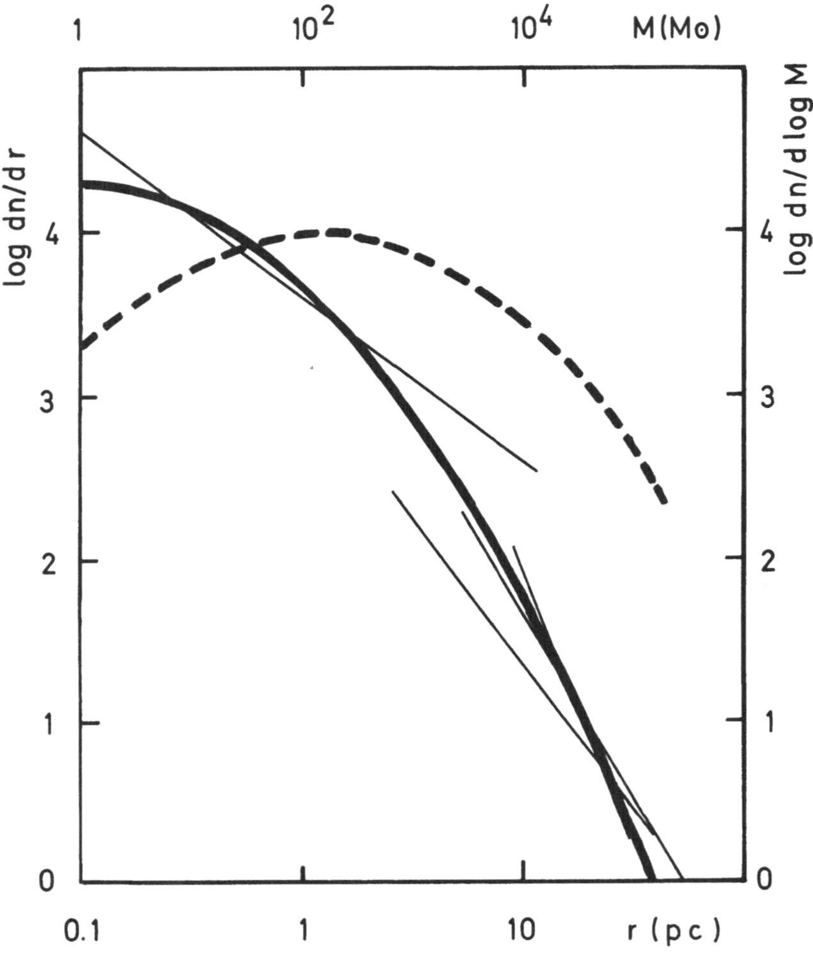

Figure 1. Left scale: Molecular cloud size distribution (thick line, see equation 2) as obtained by combining several observational results (thin lines, see equation 1 and Table I). Right scale: Corresponding molecule cloud mass distribution (dashed line, see equation 4).

References

Blitz, L. & Shu, F.H., 1980. Astrophys. J. **238**, 148.

Casoli, F. & Combes, F., 1982. Astr. Astrophys. **110**, 287.

Chièze, I.P. & Lazareff, B., 1980. Astr. Astrophys., **91**, 290.

Cowsik, R., Drapatz, S. & Michel, K.W., 1975. Proc. 14th International Cosmic Ray Conference, Vol. 1, p. 181, Max-Planck-Institut für extraterrestrische Physik, München, Germany.

Drapatz, S., 1983. Birth and Infancy of Stars, Proc. of the Les Houches Summer School, eds. Omont, A., Lucas, R., Reidel, Dordrecht, Holland.

Drapatz, S. & Zinnecker, H., 1984. Mon.Not.R.Astr.Soc., **210**, 11 p.

Elmegreen, B.G., 1979. Astrophys. J., **231**, 372.

Elmegreen, B.C., 1984a. Birth and Infancy of Stars, Proc. of the Les Houches Summer School, eds. Omont, A. & Lucas, R., Reidel, Dordrecht, Holland.

Elmegreen, B.G., 1984b. Protostars & Planets II., eds. Black, D.C., Mathews, M.S., Univ. of Arizona Press.

Felli, M., Churchwell, E., and Massi, M., 1984. Astr. Astrophys., **136**, 53.

Field, G.B. & Saslaw, W.S., 1965. Astrophys. J., **142**, 568.

Hausman, M.A., 1982. Astrophys. J., **261**, 532.

Kwan, I., 1979. Astrophys. J., **229**, 567.

Kwan, I. & Valdes, F., 1983. Astrophys. J., **271**, 604.

Larson, R.B., 1981. Mon.Not.R.Astr.Soc., **194**, 809.

Liszt, H.S., Xiang, D. & Burton, W.B., 1981. Astrophys. J., **249**, 532.

Miller, G.E., & Scalo, J.M., 1979. Astrophys. J. Suppl., **41**, 513.

Mouschovias, T.C., Shu, F. & Woodward,P., 1974. Astr. Astrophys., **33**, 73.

Parker, E.N., 1966. Astrophys. J., **145**, 811.

Rowan-Robinson, M., 1979. Astrophys. J., **234**, 111.

Solomon, P.M. & Sanders, D.B., 1980. Giant Molecular Clouds in the Galaxy, eds. Solomon, P.M. & Edmunds, M.G., Pergamon Press, Oxford.

Stark, A.A., 1979. Galactic Kinematics of Molecular Clouds, PhD thesis, Princeton University.

Stark, A.A., 1983. Kinematics, Dynamics and Structure of the Milky Way, ed. Shuter, W.L.H., Reidel, Dordrecht, Holland.

Zinnecker, H., 1984. Mon.Not.R.Astr.Soc., **210**, 43.

PART II

COMPARED INTERNAL STRUCTURE OF NEARBY MOLECULAR CLOUDS

THE SMALL SCALE STRUCTURE AND CHEMISTRY OF NEARBY MOLECULAR CLOUDS

C.M. Walmsley, T.L. Wilson
Max-Planck-Institut für Radioastronomie
Auf dem Hügel 69, 5300 Bonn 1, F.R.G.

Summary

Recent work on high density condensations in regions of nearby star formation is summarized. Special attention is given to the molecular line sources in ρ Oph B, B335, L1551, and TMC1. It is concluded that the evidence for interaction of the outflows in B335 and L1551 with their immediate surroundings is marginal. There is in both L1551 and TMC1 evidence for small scale molecular abundance variations suggesting that time dependent processes are controlling the chemistry. In TMC1, the cyanoacetylene $J=2\rightarrow1$ lines appear to be optically thick implying that current models for the density distribution in the source require revision.

Introduction

Several recent developments have made it clear that "quiescent" cold dark clouds are not so quiescent as they seem. Many harbour young pre-main-sequence objects that appear to be losing mass rapidly. These objects are found in regions of high visual extinction and hence are usually most directly studied at infrared wavelengths. They are often associated with Herbig-Haro objects and optical jets which are usually interpreted as being caused by directed outflow from the embedded pre-main-sequence star. They also give rise to the "bipolar molecular flows" seen in carbon monoxide measurements where blue-shifted and red-shifted CO are often found to be symmetrically distributed around the embedded infrared object. The recent review article by Schwarz (1983) concerning Herbig-Haro objects summarises the observational data on such flows and other contributions to this volume give some more recent results. Ho and Townes (1983) discuss our present knowledge of the gas condensations typically found in regions of star formation. In this contribution, we wish to focus upon the interaction between young newly formed stars and the surrounding gas. We will also discuss one case (TMC1) where no embedded star is known to be present.

One would in principle like to determine the physical parameters of such high density clumps and one also would like to know what influence one generation of pre-main-sequence stars has upon the next. The cold clumps, which we observe in molecular lines, probably represents the next generation of pre-main sequence stars. It is possible that such clumps are generated and maintained in motion by the action of the winds from embedded young stars (Norman and Silk (1981)). If this is the case, one might expect to see the largest linewidths and velocity-gradients in the clumps closest to the embedded star. One might also see some influence of the star upon the kinetic temperature of gas in its neighbourhood due to radiative heating of the associated dust. Finally, there is reason to believe that molecular abundances close to embedded stars could differ from those in undisturbed material. The timescale for reaching equilibrium abundances is of the same order as dynamical timescales and passage of a shock, for example, could cause the abundances of many species to be far from their equilibrium values.

One approach to these problems is to compare the properties of clumps with and without known embedded stars. Myers (this volume) has made use of recent IRAS results to study the correlation of linewidths in high density clumps with the presence or absence of infrared sources. Here, we wish to discuss a few cases in more detail with emphasis on looking for parameter variations as a function of position in the source. These cases vary from ρ Oph B, which is associated with a young star cluster, to TMC1, which (to the best of our knowledge) has no associated embedded star. We also will discuss what is known about molecular abundances in these sources. There is evidence for abundance variations within some sources over distances smaller than a parsec and such variations, if confirmed, are an important clue to molecular formation processes.

ρ Oph B

The high density gas concentration in ρ Oph B is one of four known regions where the 2 cm $2_{11}-2_{12}$ transition of H_2CO is known to be in emission. This is thought to indicate densities as high as 10^6 cm^{-3} or more (Martin-Pintado et al. (1983), Loren et al. (1983)). The formaldehyde emission region is on the flank of a larger scale molecular cloud (Wilking and Lada (1983) for which one can deduce from $C^{18}O$ measurements a mean density of $\sim 10^4$ cm^{-3} and a size of approximately 1 pc. A large number of embedded infrared objects has been found within this cloud and the ratio of stellar to gas mass may be as high as fifty percent.

Both 2 cm formaldehyde and ammonia have been mapped with the 100-m telescope over the region (Martin-Pintado et al. (1983), Zeng et al. (1984)). Figure 1 shows a comparison between the NH_3 and H_2CO distributions. On the whole, the agreement is good although ammonia does not extend so far to the north-west. While 2 cm H_2CO in emission is a density indicator, the brightness temperature of NH_3 (1,1) emission is a function of both density and ammonia column density. It is therefore by no means obvious that ammonia maps reflect the true density distribution in molecular clouds. An

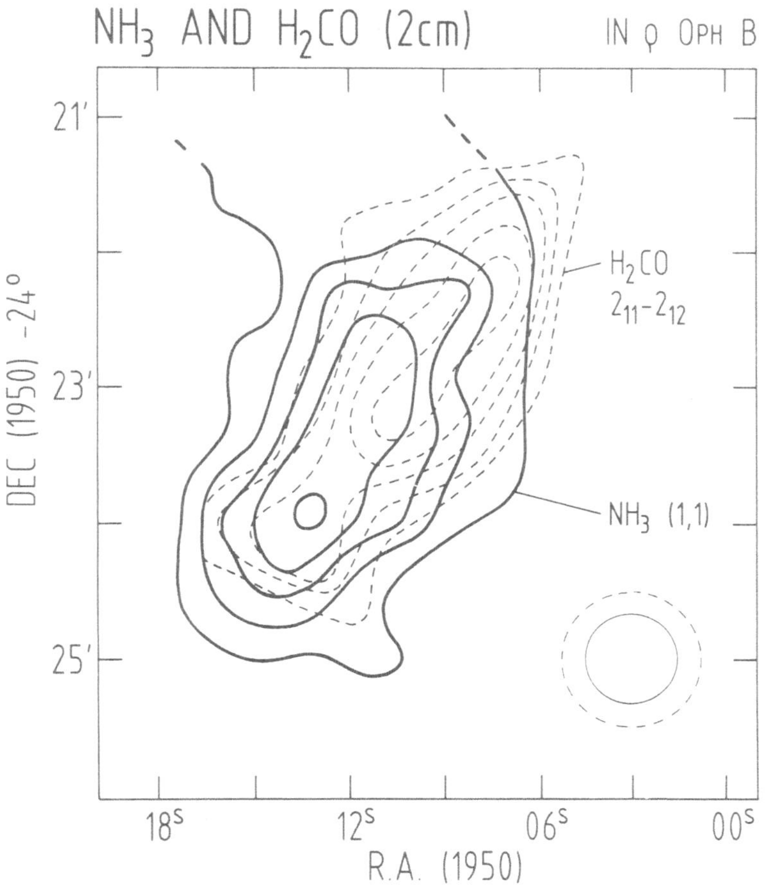

Fig. 1. *A superposition of the contours of peak $NH_3(1,1)$ brightness temperature in ρ Oph B (full lines: Zeng et al. (1984)) with contours of formaldehyde 2 cm emission (dashed lines: Martin-Pintado et al. (1983)). The circles in the bottom - right corner denote the respective half-power beamwidths.*

underabundance of ammonia in the high density material, due perhaps to depletion onto grain surfaces, could cause brightness temperature maps to be biased towards more widespread lower density gas. However, in ρ Oph B, this appears not to be the case and Zeng et al. (1984) derive an NH_3 abundance of $3\ 10^{-8}$ for the high density clump.

The temperature derived from the ammonia measurements is 18 K and is essentially constant over the region shown in Figure 1. This is much larger than the value (10 K) expected due to cosmic ray ionization and suggests that heating by young embedded stars is important. The densities derived from the formaldehyde measurements are sufficiently high that collisions with hot dust particles can efficiently heat the gas. It is reasonable to expect the gas temperature under such conditions to be close to the dust temperature (Krügel and Walmsley (1984)).

B335 and L1551

B335 and L1551 have in common that they both harbour infrared sources from which emanate bipolar outflows observed in the J=1→0 and 2→1 transitions of CO. Both have also been mapped in ammonia (Menten et al. (1984), Menten and Walmsley (1985)) and one finds that the NH_3 (1,1) peak is close to but not coincident with the embedded infrared source. The B335 infrared source is less luminous (5 $L_\odot$) and is below current sensitivity limits at wavelengths shortward of 2 μm. Its associated CO outflow (Goldsmith et al. (1984)) is oriented perpendicularly to the ammonia contours (see Figure 2). However, collimation of the flow probably takes place in a circumstellar rather than an interstellar environment (see Mundt and Fried (1983)). We do not think therefore that too much significance should be attached to the relative orientations of NH_3 and CO contours.

One fundamental question is whether brightness temperature contours of ammonia and other species are in fact representative of the molecular hydrogen distribution. In the case of L1551, this is highlighted by the differences between the Nobeyama map of J=1→0 CS (ang. res. 33") and the Effelsberg NH_3 (1,1) map (ang. res. 40") (see Kaifu et al. (1984), Menten and Walmsley (1985)). The CS map shows two peaks on either side of the embedded infrared source IRS5 suggestive of a disk oriented perpendicular to the CO outflow. The NH_3 map shows no trace of this structure although the excitation requirements of these two transitions are roughly comparable. However, column density estimates are not yet available for CS and scattering by foreground low density material may affect the observed profiles. For this reason, $C^{34}S$ measurements of the region would be useful. It seems probable to us however that real abundance variations are occurring and this clearly makes one cautious about drawing conclusions concerning the real density distribution based upon maps of the brightness temperature. It is interesting, in this context, that the ammonia and CS distributions in L183 are also highly contrasting (Ungerechts et al. (1980), Snell et al. (1982)). More recent ammonia observations by Miller (1984) show that the ammonia column density in L183 does indeed peak at the NH_3 (1,1) brightness temperature peak and is lower by almost an order of magnitude towards the CS peak. L183 has no known embedded stars and is an interesting test case for molecular formation theorists (see e.g. Millar and Freeman (1984)). However, low density foreground matter may also play an important role in this case and it is certainly true that one should understand the

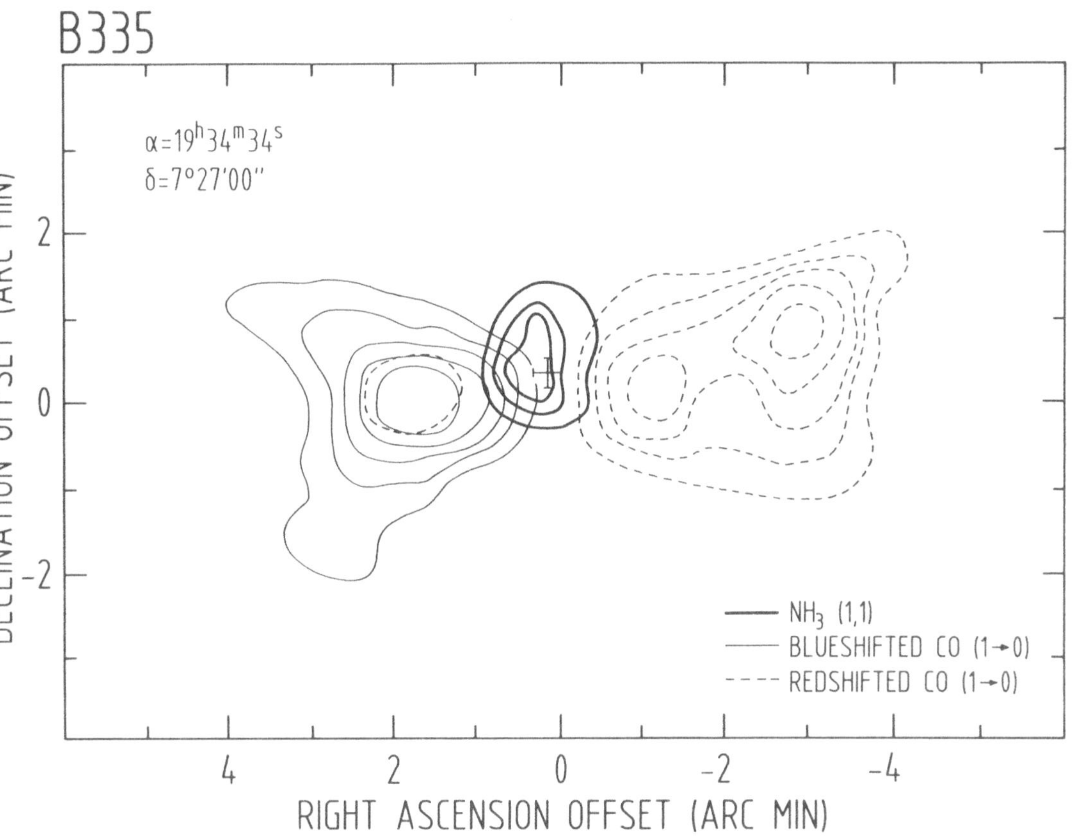

Fig. 2. *A comparison of the contours of red-shifted and blue-shifted* $J=1\rightarrow 0$ *CO emission in B335 (Goldsmith et al. (1984)) with the* NH_3 *(1,1) map of Menten et al. (1984). The cross shows the position of the far infrared source (Keene et al. (1983)).*

excitation problem before too much thought gets invested in the chemistry.

One other problem concerning molecular abundances is however worth mentioning. It is striking that there appears to be no molecular counterpart to the circumstellar dust cores found towards embedded pre-main-sequence stars by far infrared and submillimeter observations. For instance L1551 - IRS5 is found by Davison and Jaffe (1984) to be surrounded by a circumstellar shell or core of mass 0.7 $M_\odot$, temperature 35 K, and size <7 10^{16} cm (<30"). Keene et al. (1983) find in B335 a source smaller than 30" in diameter with dust temperature ∿15 K and mass ∿6 $M_\odot$. In this latter case, the H_2 column density corresponding to the angular size upper limit and the 400 μm optical depth is 10^{23} cm^{-2}. There is however no trace of these sources in the molecular line data. This could of course be due to beam dilution if the real angular size of the submillimeter sources would be much smaller than 30". The observed far infrared color temperatures however suggest that the emitting dust is relatively extended. Menten et al. (1984) have used their upper limits for ammonia emission from the dust core in B335 to derive an upper limit of 10^{-9} for the $[NH_3/H_2]$ abundance and believe that much of the molecular material in the region may have been depleted onto dust grain surfaces. It will be interesting to see whether the new generation of mm-submm telescopes can detect these compact cores in molecular lines.

A further question of interest is whether the "quiescent" gas seen in ammonia in B335 and L1551 shows signs of heating or dynamical interaction with the outflow. At least as far as heating is concerned, the answer seems to be that such interaction is small. The derived temperatures of order 10 K are similar to those found in clouds without embedded stars and can be explained by cosmic ray heating. As far as dynamical interaction is concerned, Myers (this volume) gives statistical evidence that ammonia line widths are sensitive to the presence of embedded stars. We note that the data for B335 shows no evidence for correlation of Δv with distance from the embedded star. On the other hand, in the case of L1551, it is noticeable that the NH_3 velocity gradient is largest close to the embedded star. The linewidths are smallest at the brightness temperature peak ∿30" to the SW of IRS5 and do not appear to be unusually large close to the embedded source. In fact, the peak linewidth is found in the direction of the red-shifted CO flow approximately 2' to the NW. This suggests that a study for a large number of sources of linewidths and velocities on the blue and red sides of bipolar flows would be useful. For the present, however, it is not clear whether such flows interact measurably with their surroundings. In other words, it is clear that such interaction is not large and that, perhaps due to the channeled nature of the flow, the larger part of the energy and momentum in the outflowing gas is <u>not</u> spent on accelerating the high density condensations made visible by ammonia observations. Menten and Walmsley (1984) estimate in L1551, for example, that the momentum in the high density blobs seen in NH_3 is at most 7 percent of that in the CO outflow. One implication of this is that the jet and the blobs do not directly interact.

TMC1

TMC1 differs from the other cases discussed in this article in that there appear to be no embedded stars within its high density ridge. There are however T Tauri stars in the general area of Heiles Cloud 2 and it is reasonable to look on TMC1 as a site of future low mass star formation. It is therefore of interest to understand the density structure within the cloud. TMC1 is additionally of interest because it is the site where long carbon chain molecules are particularly abundant. For this reason, accurate column density estimates for a variety of species as a function of position within the cloud are of great interest. One can also examine certain molecules in a variety of transitions. Figure 3 shows a superposition of the J=1→0 HC_3N map from Tölle et al. (1981) compared with J=5→4 and 9→8 maps from Schloerb et al. (1983). One sees in all of these maps the elongated nature of the source. One sees also that the J=5→4 and 9→8 maps are much more extended towards the north-west than the J=1→0. The reason for this has been a matter of some debate.

The J=1→0 line has a spontaneous decay rate of 4 10^{-8} s^{-1} and hence is much more easily excited than the higher J transitions which decay 100-1000 times more rapidly. This suggests that the large intensity ratios of J=5→4 and 9→8 in the north-west of TMC1 (see Bujarrabal et al. (1981)) are due to an increase in hydrogen density by a factor of around 3 between the south-eastern and north-western portions of the ridge. An alternative possibility is that saturation of the high J HC_3N lines (i.e. high optical depth) causes the maps in those lines to show relatively little contrast along the ridge. Observations of the J=2→1 transition (where the hyperfine satellites can separately be measured) show convincing evidence of high optical depths (Churchwell, Nash, and Walmsley (1984), Irvine and Schloerb (1984)). It follows that J=5→4 is also thick and (from the 1→0 results of Tölle et al. (1981)) that the HC_3N column density decreases between the cyanopolyyne (or south-eastern) peak and the north-western "ammonia" peak. By contrast, the ammonia column density increases from south-east to north-west (see Gaida et al. (1984)) and hence one can estimate that the concentration ratio $[HC_3N]/[NH_3]$ changes by a factor ∿5 over a projected distance of 0.2 pc.

Whether a simultaneous density change occurs is much less clear. Both HC_3N J=1→0 (Tölle et al. (1981) and CS data (Snell et al. (1982)) suggest that six or more fragments of typical size 0.2 pc contribute to the observed profiles. There presumably therefore exists a range of densities. Figure 4 shows the observed variation of integrated (over hyperfine satellites) line intensity towards the cyanopolyyne peak as a function of upper rotational quantum number J_u. We have derived this diagram from observed main-beam brightness temperatures using a source coupling efficiency

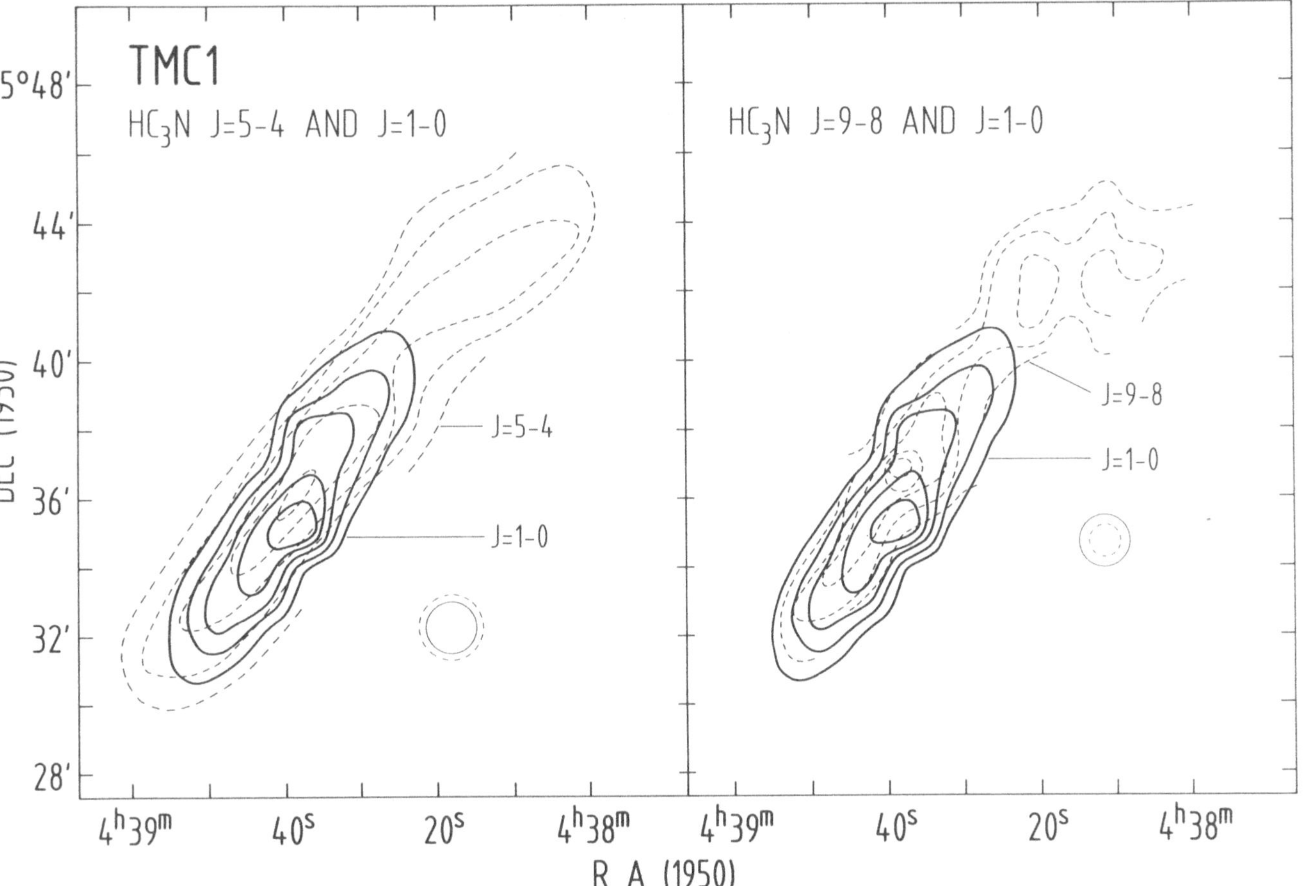

Fig. 3. *A superposition of the J=1→0 HC_3N map of TMC1 from Tölle et al. (1981) upon both the J=5→4 HC_3N (left frame) and J=9→8 HC_3N (right frame) maps of Schloerb et al. (1983).*

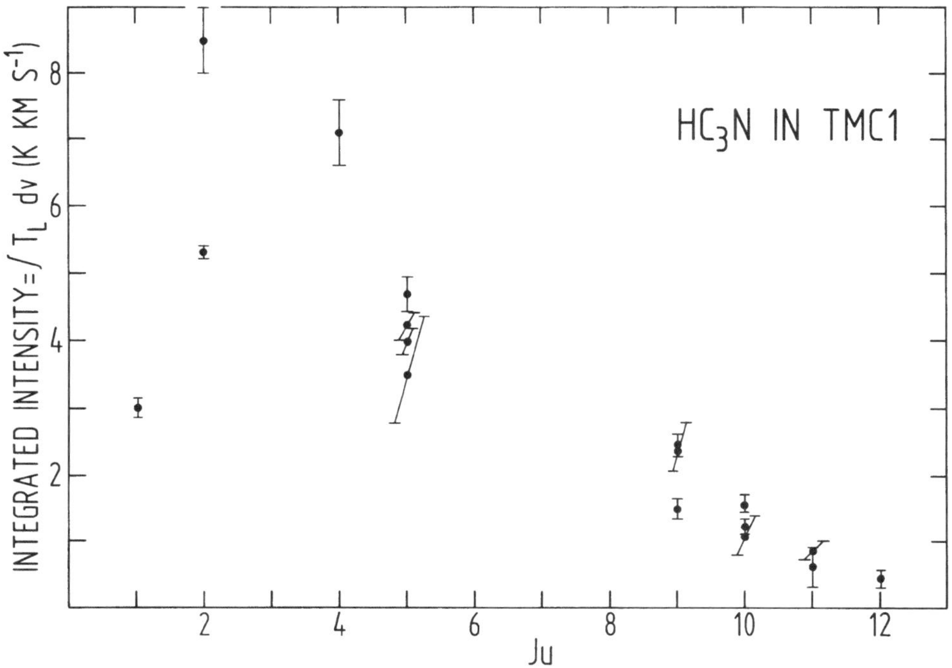

Fig. 4. *Measured values of the integrated line temperature of HC_3N rotational transitions towards the cyanopolyyne peak in TMC1 as a function of the upper rotational quantum number J_u. Data have been taken from Tölle et al. (1981), Irvine and Schloerb (1984), Churchwell et al. (1984), Avery et al. (1982), Schloerb et al. (1983), and Vanden Bout et al. (1983).*

$\eta_c = (1+\Theta_B^2/2.25)^{-0.5}$ where Θ_B is the telescope HPBW and the HPW of TMC1 perpendicular to the ridge is taken to be 1.5'. One sees from Figure 4 that the maximum occurs around J=3 and that there is evidence for a high J tail. This has been interpreted by some authors as requiring a higher excitation core and a lower excitation halo. For example, Avery et al. (1982) have put forward a model having a core with temperature = 17 K, density 6 10^4 cm^{-3}, and dimensions 1' x 2' together with a halo whose characteristics are a temperature of 7 K, density 8000 cm^{-3} and dimensions 3.5' x 10'. In contrast, Schloerb et al. (1983) model their observations with a cylinder of extent 3' x 12', a constant density of 10^5 cm^{-3}, and a temperature of 15 K. Both of these models suffer from the difficulty that the ammonia data appear inconsistent with temperatures as high as 15 K in the high density regions. They also are both essentially optically thin in all HC_3N lines and this seems inconsistent with the J=2→1 data mentioned earlier. We conclude therefore that the models presently available need revision. New attempts to discuss the situation should consider the fragments discussed by Snell et al. (1982) as distinct entities. It seems likely that the separate fragments have differing densities and abundance distributions. However, it is difficult for any model to fit the "hump" at J=2-4 visible on Figure 4.

Discussion

We have discussed three contrasting regions which have reached different stages in the star-formation process. Of these, ρ Oph B is the densest, hottest, and most massive. The enhanced temperature is almost certainly due to heating from the embedded young stars and the enhanced density could be due to stellar winds from surrounding young objects. However, in L1551 and B335 where the nature of the stellar wind is well documented, there is little evidence for the interaction of the wind with surrounding gas. It is possible that in clouds which are sufficiently massive and dense, star formation becomes self sustaining whereas in regions such as Taurus, which are less concentrated, star formation is a series of individual events. However, even in Taurus, the elongated nature of sources such as TMC1 demands an explanation. One possibility is that TMC1 is the result of a shock due to some external event which has been "fossilised" by its own self-gravitation. Another possibility is that some unspecified magnetohydrodynamic process gave rise to the apparently filamentary structure.

Finally, we note that the case for abundance anomalies and variations in molecular clouds is becoming stronger. We have discussed in some detail the case of TMC1 where $[HC_3N/NH_3]$ varies by a large factor over a projected distance of 0.2 pc. This makes it seem plausible that, for example, the observed high cyanopolyyne concentration is a "passing phase" and that the time development of the chemistry should be studied (see Williams,this volume).If so, the fragments of highest density will presumably have the shortest time scales and it may be profitable to study the relative cyanopolyyne concentrations in individual fragments (see Stahler (1984)). The ρ Oph B region with its very high density is not, incidentally, rich in cyanopolyynes. It is of course surprising that one sees any molecules at densities as high as 10^6 cm^{-3} and at dust temperatures which are not particularly elevated. One solution to this (Boland and de Jong (1982), Williams and Hartquist (1984)) is to cycle the molecular gas through regions where, due to photo-processes or shocks, the dust grain mantles are removed and the heavy atoms are returned to the gas. Such models lead to large gradients in molecular abundances over relatively short distances. They also suggest the existence of regions where molecules are absent and which can only be detected due to their dust emission. The circumstellar shells surrounding the infrared sources in B335 and L1551 may be the first examples of such regions.

References

Avery, L.W., MacLeod, J.M., Broten, N.W.: 1982, Astrophys. J. 254, 116
Boland, W., de Jong, T.: 1982, Astrophys. J. 261, 110
Bujarrabal, V., Guélin, M., Morris, M., Thaddeus, P.: 1981, Astron. Astrophys. 99, 239
Churchwell, E., Nash, A., Walmsley, C.M.: 1984, Astrophys. J. (in press)
Davidson, J.A., Jaffe, D.T.: 1984, Astrophys. J. 277, L13
Gaida, M., Ungerechts, H., Winnewisser, G.: 1984, Astron. Astrophys. 137, 17
Goldsmith, P.F., Snell, R.L., Hemeon-Heyer, M., Langer, W.D.: 1984, Astrophys. J. (in press)
Ho, P.T.P., Townes, C.H.: 1983, Ann. Rev. Astron. Astrophys. 21, 239
Irvine, W.M., Schloerb, F.D.: 1984, Astrophys. J. 282, 516
Kaifu, N., Suzuki, S., Hasegawa, T., Morimoto, M., Inantani, J., Nagane, K., Miyazawa, K., Chikada, Y., Kanzawa, T., Akabane, K.: 1984, Astron. Astrophys. 134, 7
Keene, J., Davidson, J., Harper, D.A., Hildebrand, R.H., Jaffe, D., Loewenstein, R.F., Low, F.J., Pernice, R.: 1983, Astrophys. J. Letters 274, L43
Krügel, E., Walmsley, C.M.: 1984, Astron. Astrophys. 130, 5
Loren, R.B., Sandquist, A., Wootten, A.: 1983, Astrophys. J. 270, 620
Martin-Pintado, J., Wilson, T.L., Gardner, F.F., Henkel, C.: 1983, Astron. Astrophys. 117, 145
Menten, K.M., Walmsley, C.M., Krügel, E., Ungerechts, H.: 1984, Astron. Astrophys. 137, 108
Menten, K.M., Walmsley, C.M.: 1985, Astron. Astrophys. (in press)
Millar, T.J., Freeman, A.: 1984, Mon. Not. R. Astr. Soc. 207, 425
Miller, W.: 1984, Doctoral Thesis, Univ. Köln
Mundt, R., Fried, J.: 1983, Astrophys. J. 274, L83
Norman, C., Silk, J.: 1980, Astrophys. J. 238, 158
Schloerb, F.P., Snell, R.L., Young, J.S.: 1983, Astrophys. J. 267, 163
Schwarz, R.D.: 1983, Ann. Rev. Astron. Astrophys. 21, 209
Snell, R.L., Langer, W.D., Frerking, M.A.: 1982, Astrophys. J. 255, 149
Stahler, S.W.: 1984, Astrophys. J. 281, 209
Tölle, F., Ungerechts, H., Walmsley, C.M., Winnewisser, G., Churchwell, E.: 1981, Astron. Astrophys. 95, 143
Ungerechts, H., Walmsley, C.M., Winnewisser, G.: 1980, Astron. Astrophys. 88, 259
Vanden Bout, P.A., Lorenz, R.B., Snell, R.L., Wootten, A.: 1983, Astrophys. J. 271, 161
Wilking, B.A., Lada, C.J.: 1983, Astrophys. J. 274, 698
Williams, D.A., Hartquist, T.W.: 1984, Mon. Not. R. Astr. Soc. 210, 141
Zeng, Q., Batrla, W., Wilson, T.L.: 1984, Astron. Astrophys. (in press)

DISSIPATION OF KINETIC ENERGY IN CLUMPY MAGNETIC CLOUDS

B.G. Elmegreen
IBM Thomas J. Watson Research Center
P.O. Box 218, Yorktown Heights, N.Y. 10598

ABSTRACT - The dissipation of the kinetic energy of magnetically connected clumps in a molecular cloud is discussed. For low interclump densities, dissipation is by ion-neutral viscosity inside the clumps and in the interclump medium. Additional kinetic energy is lost by Alfvén wave propagation to the external medium. The internal dissipation is found to be slow enough that molecular clouds can remain cold in the presence of supersonic clump motions. Most dissipation is by wave propagation. In the absence of wave energy input from internal or external sources, the time scale for significant energy loss from a young giant molecular cloud is approximately 2 free fall times, and in a dense molecular cloud core it is approximately 15 free fall times.

I. INTRODUCTION

Molecular clouds are likely to be composed of clumps with a variety of sizes, surrounded by a lower-density interclump medium (Zuckerman and Evans 1974). Young molecular clouds should contain the diffuse cloud-intercloud structure that was present in the ambient medium before the cloud formed. Old molecular clouds may contain clumps that are made by the disruption of shells around embedded windy stars (Norman and Silk 1980). All clouds much larger than a thermal pressure scale length should also contain clumps of a size equal to this length, which is approximately the Jeans length. Such "thermal" clumps may form by small-scale Jeans instabilities causing the gas to slip along the field until a local pressure gradient balances the clump's gravity. Cloud clumps should be interconnected by magnetic field lines, and clump-clump interactions should be moderated by this field, which stretches and gets tangled as the clumps move around. Thus, the internal energy of a cloud continously changes between kinetic energy of clump motions and magnetic energy of stretched field lines.

A net dissipation of this energy results from ion-neutral collisional viscosity. Clumps that come to a stop by the pull of a field line have their neutral particles move out ahead of their charged particles (which are connected to the field). There is a continuously changing displacement between the ions and the neutrals, and so there is a continuous frictional force between these two components. This force irreversibly converts the energy of clump motions and stretched magnetic field lines into thermal energy of the gas. When this excess thermal energy is radiated away, there is a net loss of energy from the cloud. Clump kinetic energy is also lost when the moving magnetic field lines propagate their wave energy into the external medium. This causes the external gas to become agitated, and so transfers energy of internal motion directly into energy of external motion.

Both internal and external dissipations of clump kinetic energy can limit the lifetime of a molecular cloud. Longer lifetimes require internal energy sources, such as winds from embedded stars, or external energy sources, such as supernovae, expanding HII regions, etc., which may propagate energy back into the cloud. The internal evolution of a cloud may depend somewhat on the cloud's environment.

I would like to discuss here the results of a more detailed calculation of clump energy dissipation (Elmegreen 1985 - Paper I). The full paper also considers bulk magnetic diffusion, nonlinear wave steepening, a comparison to the case of clump interactions without magnetic fields, and turbulent-type motions in a cloud. Here, the internal dissipation from interactions between clump pairs (Section II), and the propagation of internal energy to the external medium (Section III), are reviewed. Two models of cloud clumps are then considered in Section IV.

II. DISSIPATION OF INTERNAL ENERGY BY CLUMP-CLUMP INTERACTIONS

The dissipation rate of clump kinetic energy equals the clump collision rate per unit volume multiplied by the energy dissipated per collision. If the relative velocity between clumps is υ, the mean free path for clump-clump collisions (see below) is L, and the clump density in a cloud is n_c , then the collision rate per unit volume is $n_c\upsilon/L$. The energy dissipated per collision equals the ion-neutral viscous force on the clump, $N_iM_c\sigma_{in}c\Delta\upsilon$ (for ion density N_i, clump mass M_c, ion-neutral collision cross section σ_{in}, mean thermal speed c , and ion-neutral drift velocity $\Delta\upsilon$), multiplied by the net displacement between the ions and the neutrals, $2\Delta\upsilon L/\upsilon$. The ion-neutral drift velocity may be obtained by equating the ion-neutral viscous force to the inertial force on the clumps, which is $M_ca_c = M_c(\upsilon/2)^2/L$, for acceleration a_c . The factor of 2 in the velocity comes from our assumption that a clump begins its motion with velocity υ relative to another clump, and that at the time of maximum interaction (zero relative velocity) both clumps have an absolute velocity $\upsilon/2$ (to conserve momentum). The clump energy dissipation rate is therefore

$$\Lambda_B = (2n_c/\sigma_{in}c) \bullet (M_c a_c^2/N_{i,c} + M_{ic} a_{ic}^2/N_{i,ic}), \tag{1}$$

where the subscripts c and ic represent clump and interclump phases, respectively, and M_{ic} is the amount of interclump mass that can be associated with each clump.

The magnetic collisional mean free path was derived by Clifford and Elmegreen (1983):

$$L = \frac{1}{n_c \pi R_c^2} \bullet \left[\frac{4\upsilon\rho}{B_c \rho_c^{1/2}} \right]^{2/3}, \tag{2}$$

for clump radius R_c, clump density ρ_c, clump field strength B_c , and mean cloud density, ρ. This is the mean free path for stopping a test clump in a field of similar clumps, given that the clumps may interact by direct collisions or by remote entanglements of the interconnecting magnetic field lines. Equation (2) is valid only when the interclump density is low, so the interclump Alfvén velocity greatly exceeds υ, and the interclump field lines are relatively straight. For a clump field strength in pressure equilibrium with the self-gravitational pressure of the cloud, $L \sim n_c^{-1/3}$, so each clump is confined to a little cellular volume by magnetic interactions with its next-nearest neighbors. Then L is nearly independent of the interclump density.

Equations (1) and (2) should be correct for low interclump density, when the interclump field lines are nearly straight and the distortions to the field occur only at clumps, or at points where the flux tubes intersect. Then there is no field strength gradient except at these points, and no tendency for the perturbed field to steepen in the manner discussed by Zweibel and Josafatsson (1983). At the points where the field changes directions, the dissipation depends on the instantaneous acceleration. Equation (1) assumes that the acceleration and deceleration is smooth, as in a sinusoidal wave. Other possible accelerations have not been considered, and may give different results. For example, high-frequency waves may be superposed on the low-frequency field motions discussed here, and because the damping rate increases with frequency, these other modes may dissipate some of the clump energy slightly faster. If the interclump medium is dense, then the field lines between the clumps will be curved because field-line distortions produced by entanglements will propagate slowly. Then the wave that propagates the distortion may steepen by non-linear effects and dissipate more rapidly than what is calculated here.

The clump and interclump parameters in equation (1) may be determined as follows. The interclump acceleration is proportional to the clump acceleration because the interclump medium follows the magnetically attached clumps around as the clumps move in their L^3 cube. For a relatively low interclump density, the field lines that connect each clump to its nearest magnetic entanglement will be straight, and the interclump medium will move with a linear shear at a mean-squared acceleration equal to one-third of the mean squared clump acceleration (as may be derived using similar triangles-- see Paper I). Thus $a_{ic}^2 \simeq a_c^2/3$. The interclump mass associated with each clump is assumed to be proportional to the interclump density, ρ_{ic} ,

$$M_{ic} = \frac{\rho}{n_c} - M_c = \frac{\rho_{ic}}{n_c}. \tag{3}$$

The clump and interclump charged particle densities are assumed to be proportional to the square roots of the local densities, from the theory of cosmic ray ionization in molecular clouds including charge exchange, dissociative recombination, radiative recombination and recombination on grain surfaces (Oppenheimer and Dalgarno 1974; Elmegreen 1979); thus,

$$N_{i,c}/N_{i,ic} = (\rho_c/\rho_{ic})^{1/2}. \tag{4}$$

Finally, the ratio of the total mass in the interclump medium to the total mass in clumps is defined to be

$$C \equiv \frac{M_{ic}}{M_c} = \frac{\rho_{ic}}{n_c M_c}, \tag{5}$$

so the relative contribution to equation (1) from the interclump medium becomes

$$\frac{M_{ic}a_{ic}^2/N_{ic}}{M_c a_c^2/N_c} = \frac{1}{3}[C(1 + C)\rho_c/\rho]^{1/2} \equiv I - 1. \tag{6}$$

With these definitions,

$$\begin{aligned} \Lambda_B &= 0.029\left\{\frac{m_o I}{x_{i,c}\sigma_{in}c(C + 1)}\right\} \cdot \left\{\frac{v^8 B^4 \rho^5}{\rho_c^5 M_c^2}\right\}^{1/3} erg\ cm^{-3}\ s^{-1}\ (c.g.s.) \\ &= 7 \times 10^{-27}\left\{\frac{I}{C + 1}\right\} \cdot \left\{\frac{v^8 B^4 N^5}{x_7^3 N_c^5 M_c^2}\right\}^{1/3} erg\ cm^{-3}\ s^{-1}\ (a.u.), \end{aligned} \tag{7}$$

for ionization fraction $x_{i,c} = N_{i,c}m_o/\rho_c = 10^{-7}x_7$ with mean molecular mass m_o. The first expression in equation (7) is for variables in cgs units, and the second is for astrophysical units, namely, υ in km s^{-1}, B in microGauss, ρ and ρ_c in molecules per cubic centimeter (N and N_c), and M_c in solar masses; $\sigma_{in}c = 2.2 \times 10^{-9} cm^3\, s^{-1}$ from Spitzer (1978).

The internal energy dissipation time is the internal kinetic energy density, $\rho(\upsilon/\sqrt{2}\,)^2/2$, divided by Λ_B. The factor of $\sqrt{2}$ in the velocity is from our definition of υ as a root mean squared relative velocity between clumps, and not the root mean squared velocity of clumps relative to the cloud centroid. Thus, the dissipation time is

$$\tau = 4.5 \times 10^4 \left\{\frac{C+1}{I}\right\} \cdot \left\{\frac{x_7^3 N_c^5 M_c^5}{\upsilon^2 B^4 N^2}\right\}^{1/3} years, \tag{8}$$

for astrophysical units. Recall that this expression is valid only for low interclump densities, or for C<1 and I≃1.

III. ENERGY LOSS IN THE EXTERNAL MEDIUM BY ALFVEN WAVE PROPAGATION

Clump motions inside a cloud will cause the connecting magnetic field lines to move, and this field motion will propagate out of the cloud and cause the external, lower density gas to move. Thus, the internal kinetic energy of a cloud can be transferred to kinetic energy in the external medium.

The time scale, T, for this transfer of energy is approximately 1/2 of the time for Alfvén waves in the external medium to propagate out on both sides of the cloud to a distance where the waves include a total mass equal to the cloud mass (Paper I). Thus,

$$T = M/(4\pi R^2 \upsilon_{Ae}\rho_e), \tag{9}$$

for cloud mass M, radius R, external Alfvén speed $\upsilon_{Ae} = B/(4\pi\rho_e)^{1/2}$, and external density ρ_e. In astrophysical units,

$$T = 9.5 \times 10^5 \frac{M}{R^2 B N_e^{1/2}}\, years, \tag{10}$$

where N_e is the molecular density in the external medium. This external dissipation time is essentially the time scale for an Alfvén wave to propagate inside the cloud (which equals the cloud free fall time for a pressure-equilibrium field), multiplied by the square root of the ratio between the average density in the cloud and the average density in the external medium.

IV. ENERGY DISSIPATION TIMES FOR TWO CLUMP MODELS

Two examples of dissipation times for internal motions in molecular clouds are given here. One type of clump structure is likely to arise during cloud formation. Giant molecular clouds may form by the coagulation of numerous (~100) smaller clouds, each similar to the "standard" diffuse cloud found in the solar neighborhood. This coagulation may be driven by random collisions, or by the gravitational or magnetic attractions between the smaller clouds. Such "remnant diffuse clouds" (RDC) should carry their magnetic field and the connected intercloud medium along with them as they collect together. They should be in pressure equilibrium with the self-gravitational pressure of the whole cloud, and their temperatures should be around 10K, the standard temperature of quiescent molecular material in the solar neighborhood.

To calculate the energy dissipation rate for such clump structure, we assume that a giant molecular cloud has a total mass of $10^5\,M_\odot$ and a mean density of 10 cm^{-3}, and that RDC clumps each have a mass of $10^3 M_\odot$, and a density given by pressure equilibrium,

$$N_c = \frac{GM\rho}{2RkT_c}. \tag{11}$$

This pressure is calculated by assuming that the giant molecular cloud has an isothermal distribution of density (Paper I). The field strength inside each clump is also assumed to be given by pressure equilibrium,

$$B_c = (4\pi GM\rho/R)^{1/2}. \tag{12}$$

The relative velocity between the clumps is taken to be $\sqrt{2}$ multiplied by the virial theorem velocity in the cloud,

$$v = (3GM/R)^{1/2} \tag{13}$$

(again for an isothermal cloud). The values of these parameters are summarized in Table 1.

Table 1: MODEL CLOUD PROPERTIES

	Remnant Diffuse Clouds	Thermal Clumps
Surrounding Cloud:		
M, Mass ($M_\odot$)	10^5	10^3
N, Particle Density (cm^{-3})	10	10^3
R, Radius (parsecs)	35	1.6
Cloud Clump:		
M_c, Mass ($M_\odot$)	10^3	1
N_c, Particle Density (cm^{-3})	1800	38000
R_c, Radius (parsecs)	1.3	0.048
v, rms relative speed ($km\ s^{-1}$)	6.1	2.8
B_c, Magnetic field strength (μGauss)	8	36
Dissipation Results (C=0.1, x_7 = 1, $N_e = 1cm^{-3}$)		
$\Lambda_B(ergs\ cm^{-3}\ s^{-1})$	5.3×10^{-29}	4.5×10^{-26}
τ ($\times$ 10^6 years)	2100	54
T ($\times$ 10^6 years)	9.6	9.6
t_{total}($\times$ 10^6 years)	9.6	8.2
t_{total}/t_{ff}	1.7	15

The second type of cloud structure is the Jeans-scale fragmentation expected inside a strongly self-gravitating, but cold, molecular cloud core. The gas in such a core should slip along the field lines because of its gravitational self-attraction, until it establishes a local pressure gradient that can resist this attraction. The mass of the resulting condensation will be a Jeans mass or less, and the size will be the thermal pressure scale length in the cloud core, which is approximately the same as the Jeans length and much smaller than the cloud radius. We refer to these fragments as "thermal clumps" (TC). A TC should be stabilized by its internal pressure gradient until it eventually accretes so much mass from the interclump medium that it exceeds the Jeans mass and begins to collapse. The collapse presumably leads to star formation. We assume that a typical cloud core has a mass of 10^3 $M_\odot$ and a mean density of $10^3 cm^{-3}$, and that the TC mass equals 1 $M_\odot$. As for RDC's, the TC's and their magnetic fields are assumed to be in pressure equilibrium with the surrounding isothermal cloud core.

The lower part of Table 1 lists the resulting energy dissipation rates, Λ_B, for C=0.1 , $x_7 = 1$ and $N_e = 1cm^{-3}$. The internal and external dissipation times, τ and T, are also given, as is the net dissipation time obtained from the inverse sum of the two dissipation rates, i.e., from the expression,

$$t_{total} = (1/\tau + 1/T)^{-1}. \tag{14}$$

The ratio of this total energy dissipation time to the gravitational free fall time of the cloud or cloud core, $t_{ff} = (4\pi G\rho)^{-1/2}$, is found to equal ~1.7 for remnant diffuse clouds, and ~15 for thermal clumps.

An important implication of the large values of τ is that the kinetic energy of clump motions may not significantly heat a molecular cloud. The energy loss rate, Λ_B, which corresponds to the rate of transfer of kinetic energy into thermal energy inside a cloud, is much less than the gaseous cooling rate. Thus, supersonic turbulence may exist inside a cloud without much local dissipation and heating, especially if the clump/interclump density contrast is large. Furthermore, the fact that $T<\tau$ implies that most of the cloud's internal kinetic energy is lost by Alfvén wave radiation into the external medium. This loss would seem to limit the lifetime of a cloud or cloud core to only several free fall times. Longer cloud lifetimes may be possible, however, if some of this lost energy is replaced by incoming waves or by sources of wave energy inside the cloud.

REFERENCES

Clifford, P. and Elmegreen, B.G. 1983, M.N.R.A.S., 202, 629.

Elmegreen, B.G. 1979, Astrophys.J., 232, 729.

Elmegreen, B.G. 1985, Astrophys.J, submitted.

Norman, C. and Silk, J. 1980, Astrophys.J., 238, 158.

Oppenheimer, M. and Dalgarno, A. 1974, Astrophys.J., 192, 29.

Spitzer, L. Jr. 1978, in Physical Processes in the Interstellar Medium, (New York: Interscience).

Zuckerman, B. and Evans, N.J. 1974, Astrophys.J.(Lett.), 192, L149.

Zweibel, E.G. and Josafatsson, K. 1983, Astrophys.J., 270, 511.

LARGE AND SMALL SCALE STRUCTURES OF MOLECULAR CLOUDS IN THE TAURUS PERSEUS COMPLEX

E. Falgarone[1,2], J.L. Puget[1,2]
1- Radioastronomie millimétrique, ENS, 24 rue Lhomond, 75005 PARIS, France
2- DEMIRM, Observatoire de Paris, 92195 Meudon Cédex, France

N.J. Evans
Astronomy Department, University of Texas, Austin TX 78712, USA

S.R. Federman
JPL, Caltech, 4800 Oak Drive, Pasadena CA91109, USA

ABSTRACT

The gaseous components studied in HI and OH absorption against the extragalactic radiosources 3C123 and 3C111 are found to be the low-density edges of condensations of a few 100 $M_{\odot}$. The densest part of one of these condensations has a structure similar to that found in other dark clouds : it is fragmented into several cores of a few $M_{\odot}$, with an orbital velocity dispersion ~ 2 kms^{-1}. In turn, the extended low-density layer, optically thick in CO (2-1), is not a common feature. It depends on the ambient UV field but the dust temperature in the core may control its existence, even more critically: the lower is the dust temperature, the less massive is the core and the more extended is the envelope around it, for a given total mass in a given external pressure.

1. INTRODUCTION

In the course of a study of CO abundance and isotopic ratios in the low CO brightness components found in the direction of extragalactic radiosources, we were induced to map large areas around these specific lines of sight to understand the extremely low ratio R = T_A*(2-1)/T_A*(1-0) observed toward several clouds and estimate the actual pathlength through the molecular gas.

We report here on the fragmented structure found in these complexes and the existence, at the edge of the condensations, of conspicuous low density layers, optically thick in 12CO (2-1). More detailed results on the mass, size, orbital and internal velocity dispersions of the condensations will be given in Pérault (1985) and on the interpretation of all the isotopic CO (1-0) and (2-1) observations in Falgarone et al. (1985).

2. OBSERVATIONS

The low angular resolution 13CO (1-0) observations were carried out in 1982 with the Bordeaux 2.5 m antenna. A description of the calibration procedure is given in Pérault et al. (1985) (PFP). Typical line peak intensities lie between 1.5K and 2K. Selected channel maps of the 1.4°□ field around 3C123 and 3°□ field around 3C111 are displayed in Falgarone et al. (1985). The channel maps show the existence of a number of resolved condensations (or clouds) defined as in PFP as local concentrations of emission in the (α, δ, v) space. The lines of sight toward the radiosources intercept the edges of two condensations centered at v = 3.5 kms^{-1} and v = 4.6 km s^{-1} in the case of 3C123, of one condensation only in the case of 3C111.

Higher angular resolution data were obtained on selected areas of a few 10'□ with the 4.9 m antenna at the Millimeter Wave Observatory (Fort Davis, Texas), in the transitions J = 1-0 and J = 2-1 of CO, 13CO and C18O.

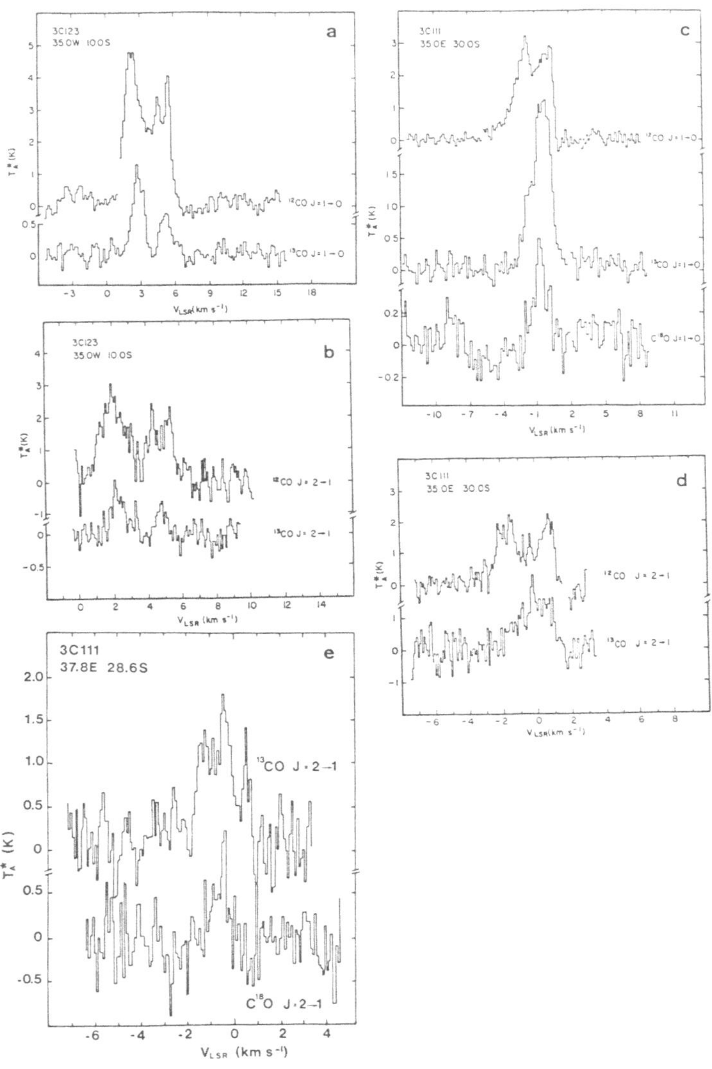

Figure 1

Preliminary CO and isotopes spectra obtained at MWO in selected directions of the 3C123 and 3C111 fields, well illustrative of the characteristic properties listed in Section 2.

The double-peaked shape of the 12CO lines is due to self-absorption in the 3C111 spectra and to the existence of two clouds on the line of sight in the 3C123 ones, the high velocity component being also clearly self-absorbed.

The 1950 reference positions are:

α =4H33M55.2S, δ =29°34.2' for 3C123

α =4H15M01.1S, δ =37°54.6' for 3C111.

A set of spectra, representative of what has been observed, is given in Fig.1 : the line of sight (35W, 10S) in the 3C123 field intercepts the low-density edges of two condensations, while those in the 3C111 field cross the densest parts of one of them. These condensations are different from those mentioned above.

The most conspicuous features of the profiles are :

i) the extremely low self-absorption dips of the 12CO (2-1) lines and the low values of the ratio R (R ~ 0.6) for the 12CO and 13CO transitions.

ii) the low ratio $R' = T_A^*(13CO)/T_A^*(C18O)$ ~ 2 (in the (2-1) transition) (Fig. 1e) found for one velocity component in the 3C111 condensation studied at small scale,

iii) the existence of several narrow velocity components (Δv ~ 0.7 kms^{-1}), marginally resolved in velocity at the position (37.8E, 28.6S) of the 3C111 field (Fig.1e), blended into one single line one beam apart (Fig. 1d).

These characteristics are not commonly found in molecular clouds observations. The third one, however, has been already noticed by PFP in another low CO brightness molecular complex.

3. LARGE SCALE STRUCTURE

The distance to the clouds is ~ 200 pc (Guibert et al., 1978) in the 3C123 field and ranges between 380 pc (Ungerer et al. (1985)) and 200pc from an estimation of the virial masses of the condensations, in the 3C111 field. They all belong to the Perseus Taurus Auriga complex mapped by Baran (1983) in the 12CO (1-0) line.

Three resolved condensations centered at v = 1.9, 3.5, 4.6 km s^{-1} can be isolated in the 3C123 (α, δ, v) channel maps. Their characteristics, derived from the 13CO (1-0) data are : diameter 1.4 to 2 pc, internal velocity dispersion (defined as $\Delta v/2.35$) 1 to 1.3 kms^{-1} and mass within the above diameters 77 $M_\odot$ to 185 $M_\odot$. The masses derived from the hypothesis of virial equilibrium are 2 times larger, in the three cases.

Two main condensations are isolated in the 3C111 field: the first, at v = 2.7 km s^{-1}, (Δv = 2 km s^{-1}, M = 2 to 7.5 10^3 $M_\odot$, D = 3.9 to 7.8 pc) has been studied by Ungerer et al. (1985), the second, at v = -0.4 kms^{-1} is centered on (35E,30S) (Δv = 1.9 km s^{-1}, M~ 590 to 2.3 10^3 $M_\odot$, D ~ 2.3 to 4.6 pc.).

4. SMALL SCALE STRUCTURE

The three characteristic features given in section 2 cannot be reproduced with a LVG radiative transfer code in a two layers structure, but appear as the signature of the whole density structure within the clouds.

We used a Monte-Carlo radiative transfer code (Bernes, 1979) in a 'core-envelope-halo' structure defined as follows: the core, (n > 10^4cm^{-3}, r < 0.06 to 0.15pc, A_v > 2mag) thermally supported by collisions of gas with dust grains, and the envelope (300cm^{-3} < n < 10^4cm^{-3}, r < 0.4pc, 1mag < A_v < 2mag) have the structure computed by Falgarone and Puget (1985) (FP) for self-gravitating condensations. In what we call the envelope, the visible extinction is significantly lower than in the terminology of Cernicharo et al. (1985). The halo (n < 300cm^{-3}, r < 2pc, A_v < 1mag) has an additional turbulent support, supposed to increase as $n^{-1/2}$. The isotopic CO abundances are those computed by Chu and Watson (1983). The simulations reproduce reasonably well the line profile temperatures observed in the direction of the (35E,30S) condensation of 3C111 but fail to mimic the actual line shapes, especially the large velocity width of the 12CO lines, which is due to the fact that the condensation develops around at least three cores scattered in velocity over ~ 2km s^{-1} This asymmetrical configuration is not taken into account in the code.

Results can be summarized as follows.

The computed line intensity ratios are more sensitive to the density structure than to any other parameter such as the isotopic ratios or the velocity fields.

The envelope-halo association is responsible for the self-absorption dip and the remarkably low temperatures in the 12CO(2-1) lines: the density gradient in the outer layers of the envelope (n ~ a few 10^3cm^{-3}) progressively smothers out, down to a few K, the emission of the optically thick core, (it is found that radiative excitation in these layers is not negligible), while the optically thick halo ($\tau \sim 1$, $T_{ex} \sim 4K$) produces the self-absorption dip. The steep density gradient of the outer envelope is therefore found to play a major role in the production of low intensity CO lines: this result shows how unreliable are the determinations of gas kinetic temperatures in clouds with such large density gradients. If this gradient is due to the rapid increase of the gas kinetic temperature in the layers where $A_v \sim 1$ (FP), then it could be a common case.

The 13CO and C18O lines characteristics are in turn a signature of the core-envelope association. Similar results have been found in another complexe and discussed in PFP.

5. COMPARISON WITH OTHER CLOUDS.

The large scale structure built from 1-3pc clouds of a few $100M_\odot$ and relative velocities of a few km s^{-1}, gravitationally bound together, is similar to that found in other complexes. The clouds of the 3C123 field have all the properties of the prominent scale of fragmentation found in molecular complexes not in the stage of forming high mass stars (PFP). In the present case, this scale is definitely resolved. The frequency of occurrence of this scale in the observations (see references in PFP) suggests that these clouds can survive in a variety of environments.

Their internal structure is in turn different from one complexe to another, as indicated by the large variety of CO and isotopes line intensities and line ratios found in the literature. It can be understood as a consequence of the high sensitivity of the line intensities to the density structure, especially to the density gradients, structure which is itself sensitive to the external radiation field (UV and IR).

The influence of the UV radiation field on the importance of the envelope relative to the core has been discussed in FP. The dust temperature in the core, controlled by the ambient IR radiation field, is also a critical parameter: the ratio of the core mass to the total mass increases as $T_d^{1.5}$ for a core-envelope structure of constant total mass, in a given external pressure and the lower is the dust temperature, the larger is the envelope. In core-envelope structures computed with T_d=15K and 90K, the line intensities are found to vary by a factor 7.5 for the C18O lines, 3.3 and 7.7 for the 13CO (1-0) and (2-1) lines and only 2.5 for the 12CO lines.

The detailed internal density structure of 1-3pc clouds as supplied by isotopic CO line observations might be considered as a sensitive indicator of the external radiation field i.e. the star formation rate in a complex.

REFERENCES

Baran, G.P.: 1983, Ph. D. Dissertation, Columbia University

Bernes, C. : 1979, Astron. Astrophys. 73, 67

Cernicharo, J., Bachiller, R., Duvert, G. : 1985, Astron. Astrophys. (in press)

Chu, Y.H., Watson, W.D.: 1983,Astrophys. J. 267,151

Erickson, N.R. : 1981, IEEE Trans. Microwave Theory Tech. Vol. MTT-29, 557

Falgarone, E., Puget, J.L. (FP): 1985, Astron. Astrophys. (in press)

Falgarone, E., Evans, N.J., Federman, S.R.: 1985, in preparation

Guibert, J., Elitzur, M., Nguyen Quang Rieu : 1978, Astron. Astrophys. 66,375

Larson, R.B.: 1981, Mon. Not. R. Ast. Soc. 194, 809

Myers, P.C., Linke, R.A., Benson, P.J. : 1983, Astrophys. J. 264, 517

Pérault, M., Falgarone, E., Puget, J.L. (PFP): 1985,Astron. Astrophys. (in press)

Pérault, M.: 1985, in preparation

Snell, R.L. : 1981, Astrophys. J. 45, 121

Ungerer, V., Mauron, N., Brillet, J., Nguyen-Quang-Rieu : 1985, Astron. Astrophys. (in press)

CHEMICAL EVOLUTION IN LOCAL INTERSTELLAR MOLECULAR CLOUDS

T.J. Millar, D.A. Williams, A.P. Jones and A.Freeman
Mathematics Department, UMIST, Manchester, M60 1QD. England

Introduction

The study of molecules in local interstellar clouds is important for several reasons. Firstly, for nearby clouds the difficulties of detecting relatively low abundances are lessened, and the opportunity exists of obtaining a fairly complete sample of molecules in a particular cloud. This is a great advantage since a wide range of observed molecules severely constrains the chemical models. Secondly, for nearby clouds the observations may show spatial variations in abundances; these variations may throw light on the formation mechanisms. Thirdly, the subject of interstellar chemistry is intrinsically worthwhile, since the models of the chemistry should - ultimately - be able to define the conditions in molecular clouds in considerable detail. For example, the models may provide information on the local radiation field, the cosmic ray flux, the relative abundance of the elements, the gas density and kinetic temperature, the level of ionization, grain optical and chemical properties, etc., (cf. Duley and Williams 1984).

Models of nearby clouds

Some of the most comprehensive observational detection lists have been obtained for nearby molecular clouds. Two of these objects, TMC1 and L 183, have been made the basis of detailed chemical modelling (Millar and Freeman 1984 a,b). These two clouds are interesting in that they are apparently similar in density and temperature yet they exhibit substantially different selections of molecules: TMC1 is much richer in the cyanopolyynes and other hydrocarbons than L 183. Millar and Freeman adopted an extensive (yet fairly conservative) gas phase network of about 500 reaction involving 120 species. The reactions were generally standard (Mitchell et al 1978, Prasad and Huntress 1980) but some particular types were expressly included to describe the formation of larger hydrocarbons and cyanopolyynes. The hydrocarbons were presumed to grow via carbon insertion and condensation reactions. H atom abstraction from, and radiative association with, H_2, and atom exchange reactions were also included. The chemistry was assumed to have attained equilibrium. By suitable choice of geometry and elemental abundances, with small variations in density, temperature, and other parameters, fairly satisfactory model fits to the observations could be obtained. In TMC1 some 13 out of 17 observed molecules were fitted by the model to within a factor of 5, the corresponding figure for L 183 being 9/13. Although such fits cannot be unique, the results suggest that detailed knowledge of these clouds, and of subtle differences between them, could be obtained from chemical modelling.

However, a notable failure occurs in the case of L 183 in that the C:CO ratio observed (Phillips and Huggins 1981) is in the range 0.1 - 1.0 whereas the value calculated from the model is $\sim 10^{-4}$. Thus, a factor of 10^3 increase in the C atom abundance is required, suggesting that the modelling is error in some fundamental way. The chemistry itself seems reasonably well understood for cool cloud equilibrium conditions.

A possible resolution of the problem could arise in the relatively slow C → CO conversion rate. At sufficiently early times, the C:CO ratio could be high. This prediction has been tested by Millar (1984) who has shown that for cloud ages between 10^5 and 10^6 years the ratio is of the correct magnitude (though it changes rapidly with time). However, the consequence of the high C abundance is that the chemistry forming hydrocarbons is driven, and overproduces hydrocarbons by several orders of magnitude, an unacceptable result. Thus, it appears that simple time-dependence cannot be the resolution of this problem, and that - since the chemistry is thought to be correct - a major physical process has been neglected.

Accretion on to grains, and its effect on cloud models

A process frequently ignored in cloud models is the collision and sticking of atoms and molecules to grain surfaces, here called accretion. Accretion of gas phase species on to grains should occur in clouds; it is a familiar process in the laboratory and is well characterized theoretically (Leitch-Devlin and Williams 1984). The time scale over which accretion is significant in molecular clouds is $\sim 10^9/n(cm^{-3})$ years; typically $\sim 10^6$ years in many clouds.

Some new evidence of interstellar accretion (and surface reaction) is available. Whittet et al (1983) have observed the 3.1μm ice band in several low extinction stars in the Taurus molecular cloud. The 3.1μm feature was previously obtained in a number of high extinction objects in which much of the available oxygen was presumed to be in H_2O. However, in the low extinction ($A_V \sim 5$ mag) clouds, most oxygen is as free atoms. Since the ice mantles exist, their presence requires the reaction $O_{gas} \rightarrow H_2O_{surface}$ to proceed fairly efficiently for mantles to build up in the available time. This constitutes the first direct evidence of surface chemistry on interstellar grains (Jones and Williams 1984). Similar reactions on the surface should also form other species (Williams 1984, Duley and Williams 1984) and the dirty ice resulting gives rise to a profile of the 3.1μm feature similar to that observed (Knacke et al 1982 Hagen et al 1983, Leger et al 1983).

The evidence for accretion on a time scale $< 10^6$ years poses the question: why are molecules observed in molecular clouds? Further studies by Millar (1984) - see also Iglesias (1977) - confirm that all heavy atoms and molecules are severely depleted by about 10^6 years. If the age of the denser clumps is less than this, then accretion is not a serious loss. Such a claim has been made for the dense knots of gas in the Taurus molecular cloud (Myers 1984); in this case accretion is unimportant and the molecular abundances must be those of young objects. Problems concerning these abundances may lie with the initial conditions adopted in the models. More massive clouds, such as that towards ρ Oph, are longer lived and accretion should be important. Since molecules are observed, in such clouds some mantle limitation process must be operating.

Processes limiting mantle growth

A number of proposals have been made. Greenberg (1982) has suggested a mechanism which depends on the triggering of reactions between radicals accumulated in the mantle; the heat generated in the runaway reaction removes the accumulated material. Radical reactions have also been discussed by Allen and Robinson (1975) and Duley (1984). Leger et al (1984) have proposed that cosmic ray heating of grains causes thermal evaporation. The internal UV radiation field arising from cosmic ray excitation of H_2 (Prasad and Tarafdar 1983) also provides heating for grains or a photodesorption source. Boland and de Jong (1982) have involved turbulent transport, with photodesorption of mantles of dredged-up grains by the ambient radiation field or by SN optical light flashes.

Williams and Hartquist (1984) have proposed that occasional low velocity shocks limit mantle growth and return adsorbed CO to the gas as atoms, so resolving the problem of the high C: CO ratio. On this view, all dense molecular material should be shocked every million years or so, otherwise becoming undetectable, owing to the effectiveness of accretion. This frequency of shocks is comparable with that required in the T Tauri modulated cloud model of Norman and Silk (1980). Prasad (1984) has also proposed to limit mantle growth by restricting the cloud age to about a million years. His model of a gravitationally collapsing cloud, truncated in age by star formation, has many attractive features.

Conclusion

We conclude that chemical models of nearby molecular clouds can give some detail about local interstellar conditions. However, the models are currently incomplete in that surface processes, leading either to accretion or to reaction, and involving some mantle limitation process, require to be included.

References

Allen, M. and Robinson, G.W. 1975. Astrophys. J. 195, 81.
Boland, W. and de Jong, T. 1982. Astrophys. J. 261, 110.
Duley, W.W. 1984. Pre-print.
Duley, W.W. and Williams, D.A. 1984. "Interstellar Chemistry", Academic Press, London.
Greenberg, J.M. 1982. "Submillimetre Wave Astronomy", eds. J.E. Beckman and J.P. Phillips, Cambridge University Press.
Hagen, W., Tielens, A.G.G.M., and Greenberg, J.M. 1983. Astr. Astrophys. 117, 132.
Iglesias, E. 1977. Astrophys. J. 218, 697.

Jones, A.P. and Williams, D.A. 1984. Mon. Not. R. astr. Soc. 209, 955.
Knacke, R.F., McCorkle, S., Puetter, R.C., Erickson, E.F. and Kratschmer, W. 1982. Astrophys. J. 260, 141.
Leger, A. Klein, J., de Cheveigne, S., Guinet, C., Defourneau, D. and Belin, M. 1979. Astr. Astrophys. 79, 256.
Leger, A., Jura, M., and Omont, A. 1984. Pre-print.
Leitch-Devlin, M.A. and Williams, D.A. 1984. Mon. Not. R. astr. Soc. 210, 577.
Millar, T.J. 1984. Paper given at the NATO Advanced Study Institute "Molecular Astrophysics - State of the Art and Future Directions", Bad Windsheim, West Germany.
Millar, T.J. and Freeman, A. 1984a. Mon. Not. R. astr. Soc. 207, 405.
1984b. ibid. 207, 425.
Mitchell, G.F., Ginsburg, J.L. and Kuntz, P.J. 1978. Astrophys. J. Suppl. 38, 39.
Myers, P.C. 1984. this volume.
Norman, C. and Silk, J. 1980. Astrophys. J. 238, 138.
Phillips, T.G. and Huggins, P.J. 1981. Astrophys. J. 251, 533.
Prasad, S.S. 1984. NATO Advanced Study Institute.
Prasad, S.S. and Huntress, W.T. 1980. Astrophys. J. Suppl. 43, 1.
Prasad, S.S. and Tarafdar, S.P. 1983. Astrophys. J. 267, 603.
Whittet, D.C.B., Bode, M.F., Longmore, A.J., Baines, D.W.T., and Evans, A. 1983. Nature 303, 218.
Williams, D.A. 1984. "Galactic and Extragalactic Infrared Spectroscopy", eds. M.F. Kessler and J.P. Phillips, D. Reidel, Dordrecht.
Williams, D.A. and Hartquist, T.W. 1984. Mon. Not. R.astr. Soc. 210, 141.

$C^{18}O$ AND OPTICAL OBSERVATIONS OF THE TAURUS CLOUD IN FRONT OF 3C 111

V. Ungerer[1], N. Mauron[2], J. Brillet[3], Nguyen-Quang-Rieu[1]
(1) Observatoire de Meudon, DEMIRM, F-92190, Meudon, France.
(2) Observatoire du Pic du Midi et de Toulouse, 14 av Edouard Belin, 31400, Toulouse, France
(3) Observatoire de Bordeaux, F-33270, Floirac, France.

I) INTRODUCTION:

Distance estimates and large scale surveys in OH (Wouterloot 1980) and ^{12}CO (Baran 1982) suggest that the molecular material in the Taurus region is a superposition of distinct molecular complexes which will be referred to as Taurus A, Taurus B and Perseus following Wouterloot (1980). Taurus A contains high density clumps such as TMC 1 and TMC 2 and is located at 140 pc (Elias 1978) and the Perseus complex is located near the Per OB2 association whose distance is ~ 350 pc. Taurus B is an elongated structure (14°x2°) parallel to the galactic plane at b ~-8°. A maser emission in the 1612 MHz OH line and an OH main line anomaly have been detected on the line of sight of 3C 111 (l^{II} = 161.7°, b^{II} = -8.8°) (Winnberg et al. 1984). We present here optical and $C^{18}O$ observations around the position of 3C 111 which allow the determination of the distance, the density and velocity structure of the cloud and the $C^{18}O/A_V$ correlation.

II) DISTANCE OF THE CLOUD:

Prism - objective spectral classification and UBV photometry of field stars in a 2°x2° region around 3C 111 have been performed with the prism objective telescope (GPO) and the 1-m Chiran telescope at the Haute-Provence Observatory. On another hand, photometric data of stars located in a large field (10°x10°) adjacent to the Taurus-Perseus complex have been taken from catalogues of the Stellar Data Center (CDS) of Stasbourg Observatory (see details in Ungerer et al. 1984). Visual extinction is plotted against distance for stars lying beside the molecular complex (Fig. 1a) and toward the molecular complex (Fig. 1b) respectively.

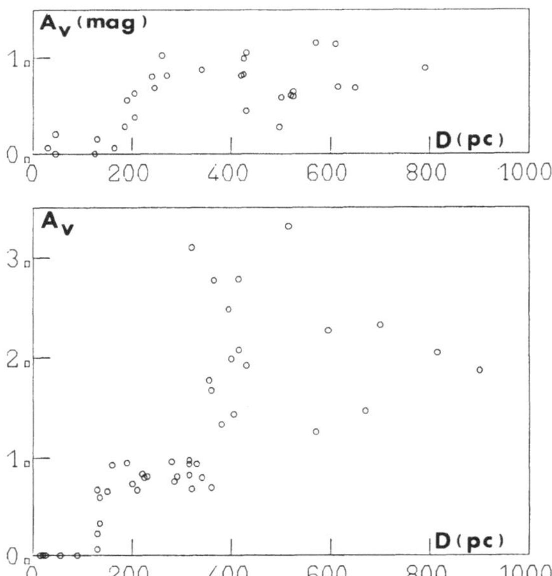

Figure 1 a, b : visual extinction versus distance for stars lying a) in a large field adjacent to the Taurus-Perseus complex (b) in a 2°x2° field aroud 3C 111.

In the direction adjacent to the molecular complex there is evidence of a diffuse absorption layer ($A_V \sim 0.9$ mag) between 140 and 240 pc (Fig. 1a). In the direction of the "molecular cloud 3C 111" we observe two distinct absorption layers: a diffuse one ($A_V \sim 0.8$ mag) at $d = 135 \pm 10$ pc and a more opaque and clumpy one at $d = 380 \pm 20$ pc (Fig. 1b).

We interpret the evidences of a diffuse layer on a large scale at 140 pc as an extended halo of the Taurus A molecular complex. At their common distance, the separation between the position of 3C111 and the densest parts of Taurus A complex is $\sim$ 30 pc. The slower increase of the visual extinction at 140 pc for the stars located beside the Taurus complex (Fig. 1a) can be explained by the large angular distribution of these stars.

The second absorption layer at $\sim$ 380 pc can be attributed to the molecular cloud associated with the Taurus B complex. It's distance is close to the distance of the Per OB2 association whose parent cloud is beleived to be the Perseus complex. The 2 molecular complexes, Taurus B and Perseus, could have been related in the past and disrupted by the formation of the Per OB2 association.

III) STRUCTURE OF THE "3C 111 CLOUD":

Star counts on the POSS plates over an area of 1.4°x1.2° around the position of 3C 111 indicate that the part of the cloud in front of 3C 111 is not the densest part ($A_V \sim 1.3$ mag) and reveal the existence of a large condensation (35' in diameter) located north-east of 3C111.

The $C^{18}O$ J = 1 - 0 emission has been mapped with the 2.5-m antenna of the Observatory of Bordeaux -France- (see Baudry et al. 1981), over the whole condensation with a grid of spacing of 5'. Details of the observational procedure as well as line profiles and characteristics are given in Ungerer et al. (1984).

Maps of $C^{18}O$ emission intensity per velocity channel (channels' width ~ 0.27 kms^{-1}) are presented in Figure 2, where the offsets are given relatively to the position of 3C111 (α(1950)= $04^H15^m01.1^s$; δ(1950)=+37°54'37")). On the first 4 pannels, we observe the evolution of 3 clumps in the central and west part of the condensation with a maximum of emission at $V_{LSR} \sim -3.25$ kms^{-1}. Then the emission drops sharply at $V_{LSR} \sim -2.70$ kms^{-1}. And an another set of clumps, adjacent to the first one, appears at $V_{LSR} \sim -2.4$ kms^{-1} with a ringlike structure. A similar velocity pattern has been interpreted in Heiles Cloud 2 as a rotating ring by Schloerb and Snell (1984). On the ^{13}CO maps (Falgarone private communication), this effect does not appear clearly although most clumps have their counterpart. The two $C^{18}O$ velocity components have their counterparts in the OH spectra toward 3C 111 taken with the Effelsberg 100-m antenna (HPBW $\sim$ 8'). They exhibit 3 velocity components at $V_{LSR} \sim -1.0$ kms^{-1}, $V_{LSR} \sim -2.4$ kms^{-1} and $V_{LSR} \sim -3.2$ kms^{-1} respectively, the third one being only marginaly detected (Winnberg et al. 1984).

For a distance of 380 pc the linear extent of the $C^{18}O$ emission is 4.2 pc. The total mass of the condensation derived from the $C^{18}O/A_V$ correlation (see below) is 650 $M_{\odot}$. Assuming a spherical geometry this leads to $n(H_2) \sim 280$ cm^{-3}.

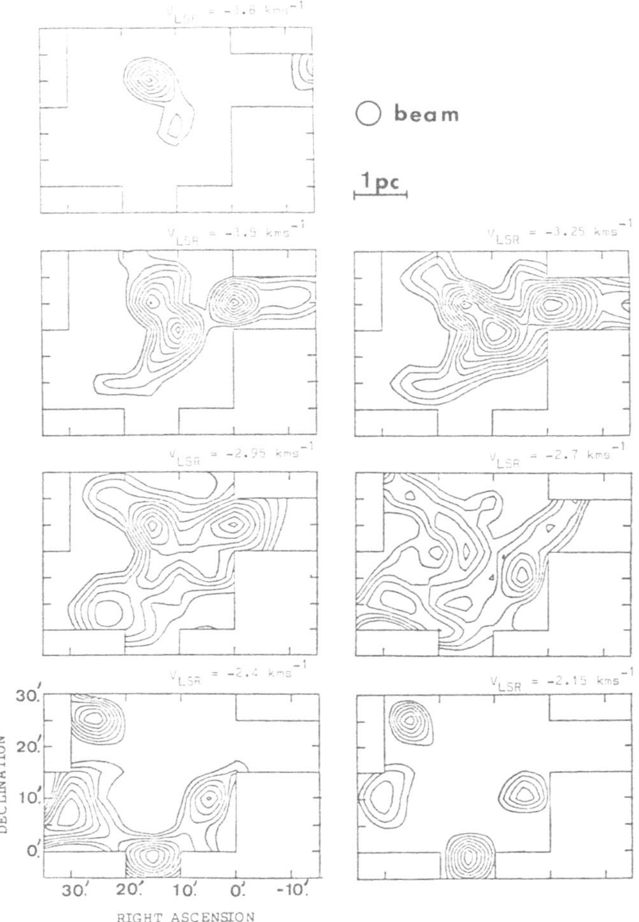

Figure 2: Maps of $C^{18}O$ intensity $T_A^*(C^{18}O$: J=1-0) per velocity channel. The (0,0) position is 3C 111 and the velocity is indicated in the right corner. The spacing of contours is 0.05 K. The first contour is 0.2 K.

IV) CORRELATION BETWEEN $C^{18}O$ COLUMN DENSITY AND VISUAL EXTINCTION:

The correlation between $N(C^{18}O)$ and A_V has been studied by Frerking et al (1982) in selected regions of Taurus A and ρOph, and found to differ strongly in these two regions and to consist of two regimes, the transition occuring at $A_V \sim 4$ mag. We have used the visual extinctions derived from our star counts and the $C^{18}O$ data to study their relation in the "3C 111 cloud" ($A_V < 4$ mag).

The $C^{18}O$ column densities have been derived as in Frerking et al. (1982), assuming a constant excitation temperature T_{ex} = 10 K (infered from unpublished ^{12}CO observations).

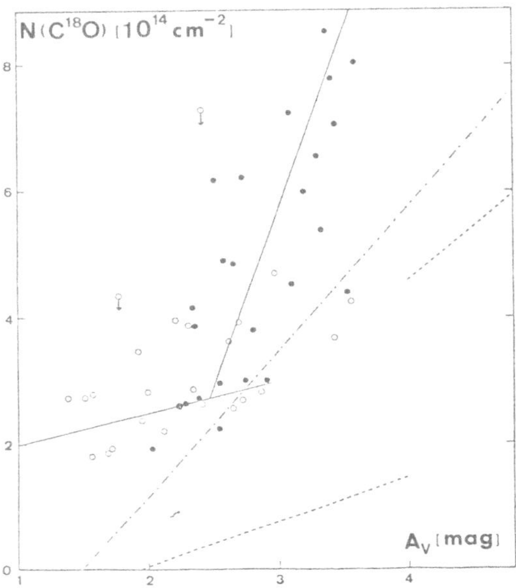

Figure 3: $C^{18}O$ column density as a function of visual extinction. Filled circles (open circles) correspond to $C^{18}O$ data with signal to noise ratio higher (lower) than 6. The dashed lines (- - -) are the relationships derived by Frerking et al (1982) in selected regions of Taurus A for $2 < A_V < 4$ and $A_V > 4$ respectively. The dashed line (-·-·-·) indicates the relationship derived by Cernicharo (1985) in Heiles Cloud 2.

The column density of $C^{18}O$ is plotted against the visual extinction in Figure 3. The full lines indicate the linear least squares fit along the 2 coordinates. The dashed lines indicate the relations determined by Frerking et al.(1982) and Cernicharo (1985) in Taurus A.

We observe a sharp turnover of the $N(C^{18}O) - A_V$ curve at $N(C^{18}O) \sim 3\ 10^{14}\ cm^{-2}$ in accord with theoretical models of selective photodissociation of $C^{18}O$ through UV line absorption. Although the transition between the two regimes occurs at the same column density as found by Frerking et al.(1982) in Taurus A and ρ Oph, they are systematically higher by at least a factor 7 in the "3C 111 cloud". This discrepency if real (Frerking et al. have only 6 points), as well as the rather normal ETL isotopic ratios we derive ($N(^{13}CO)/N(C^{18}O) \sim 5 - 13$) could be explained by an enhanced UV radiation field in the region of the Taurus A clouds and even more of the ρ Oph molecular cloud resulting in higher fractionation effects and CO isotope selective photodissociation (Chu and Watson (1983)).

REFERENCES:

Baran, G.P.: 1982, PhD. Thesis, Univ. Columbia.

Baudry, A., Cernicharo, J., Perault, M., de la Noë, J., Despois, D.: 1981, Astron.Astrophys. 104, 101.

Cernicharo, J.: 1985, this volume.

Chu, Y.H., Watson, W.D.: 1983, Astrophys. J. 267, 151.

Elias, J.H.: 1978, Astrophys.J. 224, 857.

Frerking, M.A., Langer, W.D., Wilson, R.W.: 1982, Astrophys.J. 262, 590.

Schloerb, F.P., Snell, R.L.: 1984, Astrophys.J. 283, 129.

Ungerer, V., Mauron, N., Brillet, J., Nguyen-Quang-Rieu: 1984, Astron. Astrophys. (accepted).

Winnberg, A., Nguyen-Quang-Rieu, Ungerer, V.: 1984, in preparation.

Wouterloot, J.: 1980, PhD. Thesis, Leiden.

RADIO CONTINUUM AND MOLECULAR LINE STUDIES OF DENSE CORES IN DARK CLOUDS

I.I. Zinchenko and A.G. Kislyakov
Institute of Applied Physics
Academy of Sciences of the USSR
Gorky, USSR

In recent years the star formation process in interstellar clouds has been understood in more detail. However, very little is known about the stage prior to the appearance of a new star, when a small cloud of stellar mass contracts without embedded energy sources. Search for these objects and their thorough investigation is one of the most intriguing astrophysical problems.

We have conducted for some years a program of such studies by observations of millimeter wave continuum and molecular line radio emission. The observations were made with the 22-m radio telescope of the Crimean Astrophysical Observatory, the 25 x 2-m radio telescope of the Institute of Applied Physics, and with the radio telescopes in USA, FRG and Sweden.

Continuum observations allow for detection of sufficiently dense dust condensations. Because the gas-to-dust mass ratio is nearly constant, these observations can be used in search for regions of enhanced gas density. The continuum emission intensity of the cold (5 - 10 K) dust clouds peaks at $\lambda \sim 0.5$ mm. It can be shown that under the clear sky conditions the most dense condensations (with the dust optical depth $\tau_d \gtrsim 0.05$ at $\lambda \sim 1$ mm) can be found more effectively by continuum observations than by molecular line observations (Zinchenko and Kislyakov 1979). The minimal detectable mass of condensation, if the latter is larger than the telescope beam, is

$$M_{min} = \frac{\pi}{6} \frac{D_{min}^3}{X_d^3 m^2 n^2}, \qquad (1)$$

where X_d is the dust-to-gas mass ratio (0.01), m is the hydrogen molecule mass, n is the gas density, and D_{min} is the minimal detectable dust column density. When the cloud radius achieves a value $R \leqslant R_c$, its angular size becomes less than the size of the beam. Then

$$M_{min} = \pi \frac{D_{min} R_c^2}{X_d} \tag{2}$$

The minimal antenna temperatures, which are currently detectable in dark clouds, are $\sim$ 3 mK at a wavelength $\lambda \sim 3$ mm. Then using the values of the grain absorption efficiency obtained by Sherwood, Arnold and Schultz (1980) we have $D_{min} \sim 3\ 10^{-4}$ g/cm^2. In Figure 1 estimates of minimal detectable mass are presented assuming $R_c =$ $= 0.03$ pc. They are $\sim$ $0.3\ M_\odot$.

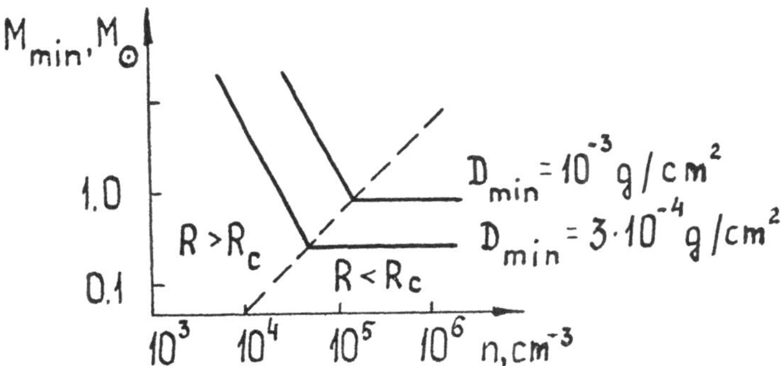

Fig. 1. The minimal detectable mass of a dust condensation ($R_c = 0.03$ pc, $X_d = 0.01$)

Condensations can be also searched for by observations of molecules which are observed in regions of high density, e.g., HCN, CS and some others. The minimal detectable masses are also $\sim$ 0.3 - - 1.0 $M_\odot$ (Padman 1982).

At the Institute of Applied Physics, 20 dark clouds have been observed in continuum at 1-4 mm and in the J=1-0 HCN (λ = 3.4 mm) line. According to the standard detection criterion ($I > 3\sigma$) the HCN line was detected in about a half of the clouds (Burov et al. 1982). Analysis shows that the gas density in regions of the line formation should be $n \gtrsim 10^5$ cm^{-3}. In some clouds the line intensity is anomalously high. Recently observations of these clouds were repeated with a better spectral resolution (350 kHz) and improved sensitivity. The preliminary results confirm the existence of high-intensity HCN emission in some of them.

In three clouds, continuum sources at λ = 1 - 4 mm were detected (Kislyakov et al. 1973, Kislyakov et al. 1976, Zinchenko et al. 1978). However, the uncertainties of the continuum data are still very high. Moreover, there are some discrepancies in these data. In L1640 the source position coincides with the position of a strong 100-μm source (Hoffman, Frederick and Emery 1971). The source spectrum is plotted in Fig. 2. It is well fitted by a function

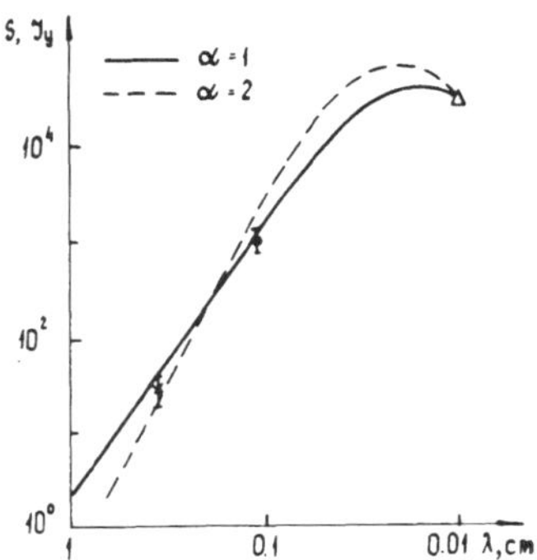

Fig. 2. The continuum spectrum of the source in the L1640 cloud. The 3-mm detection is marginal

$[1-\exp(-\tau)]B(\nu,T)$, where $B(\nu,T)$ is the Planck function, $\tau(\nu)$ is the optical depth (T = 25 K, $\tau(\nu) \propto \nu^{\alpha}$). The CO line brightness temperature increases sharply near the continuum source where it achieves 25 K (Kislyakov and Turner 1976). Therefore the gas and dust temperatures are nearly equal. A compact H II zone was revealed here by observations at the NRAO interferomater (flux density 12 mJy and size $\sim 6''$ at a wavelength of 11 cm (Kislyakov and Turner 1976). Such a zone can be excited by a B2V star. The IR luminosity corresponds to a BOV star. This picture is typical of a recently born early-type star.

The other two continuum sources are apparently colder. The CO peak antenna temperature is nearly constant and is approximately equal to 7 K, a value which corresponds to an excitation temperature of $\sim$ 10 K (Kislyakov and Gordon 1983). The main CO isotope line in L673 has an asymmetric self-reversed profile (Fig. 3). Usually, such a dip in the line profile is interpreted as a result of self-absorption in a cold expanding envelope. However, an alternative interpretation is possible. It is well known that the kinetic temperature increases toward the boundary in dark cold clouds. If the density is sufficiently high, the CO excitation temperature can also increase toward the edge. In this case, a similar profile can arise but in a contracting cloud. Numerical results obtained by the Monte-Carlo

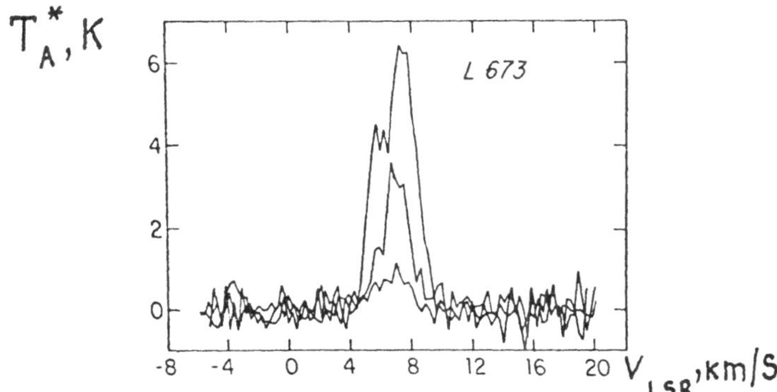

Fig. 3. The J = 1 - 0 CO, ^{13}CO and C^{18}O emission spectrum in the dark cloud L673 (Kislyakov and Gordon 1983)

method show that this is possible if the gas density in the centre is $n \gtrsim 10^5$ cm^{-3} (Zinchenko and Lapinov 1985). Examples of the calculated profiles for CO and ^{13}CO are shown in Fig. 4. In this mo-

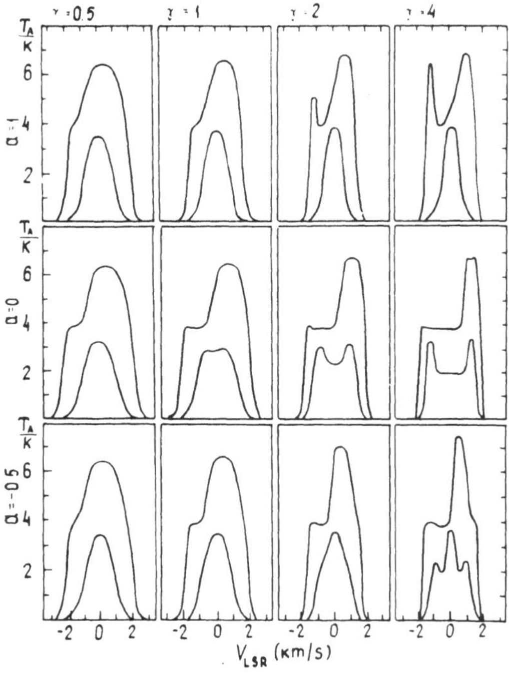

Fig. 4. CO and ^{13}CO line profiles calculated for collapsing cold core models with different velocity laws ($V \propto r^{\alpha}$, α = = -0.5, 0, 1) and systematic velocity to turbulent velocity ratios (γ)

del, the kinetic temperature is ~ 7 K in the centre and ~ 15 K at the boundary. This model explains the observed $T_A^*(CO)/T_A^*(^{13}CO)$ ratio in a natural way (the optical depth in the ^{13}CO line is probably $\tau(^{13}CO) > 1$). The variations in the CO line profile across the source do not contradict the predictions of this model but the CO data are incomplete and too noisy. The observations of NH_3 and HCN molecules do not contradict them either. In NH_3 observations the (1,1) transition was detected, while the (2,2) transition was not. This implies an upper limit of the kinetic temperature of ~ 9 K (Kislyakov et al. 1983). The HCN observations (Sandell, Höglund, Kislyakov 1983) confirm the existence of a high-density core in this cloud. The HCN optical depth is moderate ($\tau \sim 3 - 10$). Taking this into account and using the results of the HCN excitation analysis we can conclude that the H_2 density in the L673 core should be $n \gtrsim 3 \cdot 10^4$ cm^{-3}. The H_2CO observations set an upper limit on the cloud density because the 2-cm H_2CO line is seen in absorption (Kislyakov et al. 1983). This implies that $n < 10^6$ cm^{-3}. Thus, seemingly, there is a dense cold contracting core in the L673. The core mass is 300 $M_\odot$ according to the 3-mm continuum observations though this value depends on the assumed grain emissivity and distance to the source. It should be noted that the results of 1-mm measurement (G. Sandell, private communication) seem to be inconsistent with our 3-mm data though the source was not mapped at 1 mm. This picture resembles theoretical conceptions on a collapse of protostellar clouds.

The results presented demonstrate that in many dark clouds dense cores exist, some of which probably contract.

References

Burov, A.B., Voronov, V.N., Zinchenko, I.I., Krasil'nikov, A.A., and Kukina, E.P. 1982, Astron.Zh., 59, 267

Hoffman, W.F., Frederick, C.L., and Emery, R.J. 1971, Astrophys.J., 170, L89

Kislyakov, A.G., Chernyshev, V.I., Listvin, V.N., and Shvetsov, A.A. 1973, Izv. VUZ, Radiofizika, 16, 774

Kislyakov, A.G., and Turner, B.E. 1976, Astron.J., 81, 302

Kislyakov, A.G., Chernyshev, V.I., Grigoryan, F.A., and Khachatryan, P.R. 1976, Pis'ma v Astron.Zh., 2, 240

Kislyakov, A.G., and Gordon, M.A. 1983, Astrophys.J., 256, 766.

Kislyakov, A.G., Walmsley, C.M., Winnewisser, G., Batrla, B. 1983, XV All-Union Conf. on Galactic and Extragalactic Radio Astronomy, Kharkov. Abstracts, 203

Padman, R. 1982. In: The Scientific Importance of Submillimetre Observations. Proc. ESA Workshop, Noordwijkerhout, p.29

Sandell, G., Höglund, B., and Kislyakov, A.G. 1983, Astron.Astrophys., 118, 306

Sherwood, W.A., Arnold, E.M., and Schultz, C.V. 1980, In: Interstellar Molecules (ed. B.H.Andrew), Reidel, Dordrecht, p.133
Zinchenko, I.I., Kislyakov, A.G., Krasil'nikov, A.A., Kukina, E.P. 1978, Pis'ma v Astron.Zh., 4, 10
Zinchenko, I.I., and Kislyakov, A.G. 1979. In: Spectral Studies of Cosmic and Atmospheric Emission, Gorky, p.33
Zinchenko, I.I., and Lapinov, A.V. 1985, Astron.Zh., 62

STAR COUNTS AND AMMONIA OBSERVATIONS IN THE CENTRAL TAURUS REGION

H.Ungerechts, G.Winnewisser, M.Gaida
I.Physikalisches Institut
der Universität zu Köln
Zülpicher Str.77
D-5000 Köln 41, Federal Republic of Germany

We present results from ongoing work on the Taurus molecular clouds dealing with (a) the large scale structure of the dust lanes and (b) the physical characteristics of the embedded high density cloud cores.

We have obtained maps of the visual extinction A_V from star counts in 4!5*4!5 fields for the two filaments known as Kutner's cloud and Barnard's cloud as well as for two isolated groups of small dark clouds, L1489 and L1517 (Gaida et al., 1984). The angular resolution of 4!5, corresponding to 0.18pc at a distance of 135 pc (Elias, 1978), is sufficient to show fragments having sizes of 0.2pc - 1.0pc. From A_V the mass M of the fragments and the mass per length M/l of the filaments can be estimated and compared to theore tical values for equlibrium configurations (Ostriker, 1964, Bastien, 1983). This analysis suggests that the large filaments are unstable and breaking into denser fragments by contracting along their major axes.

Embedded in the regions of high A_V are the small dense cold cores which show up in molecular lines tracing gas of density $>10^4 cm^{-3}$, such as the ammonia (J,K)=(1,1)&(2,2) inversion transitions. During the last years we obtained data for these lines in 9 such cloudlets in Taurus with high spatial resolution using the Effelsberg 100m telescope (Ungerechts et al., 1982, Gaida et al., 1984, Miller, 1984). In the accompanying figure we show the maps of the (1,1) peak brightness temperature of four of these cloud cores. In general, their typical scale-size is 1'-2', slightly larger than the beam diameter (40"). It seems interesting to compare physical parameters derived from NH_3 observations at Effelsberg to those determined with the smaller Haystack- and NRAO- telescopes, with which a much larger sample of cloud cores has been studied (see, e.g., Benson and Myers, 1980, Myers and Benson, 1983, Benson and Myers, 1983, and P.Myers' contribution to this volume).

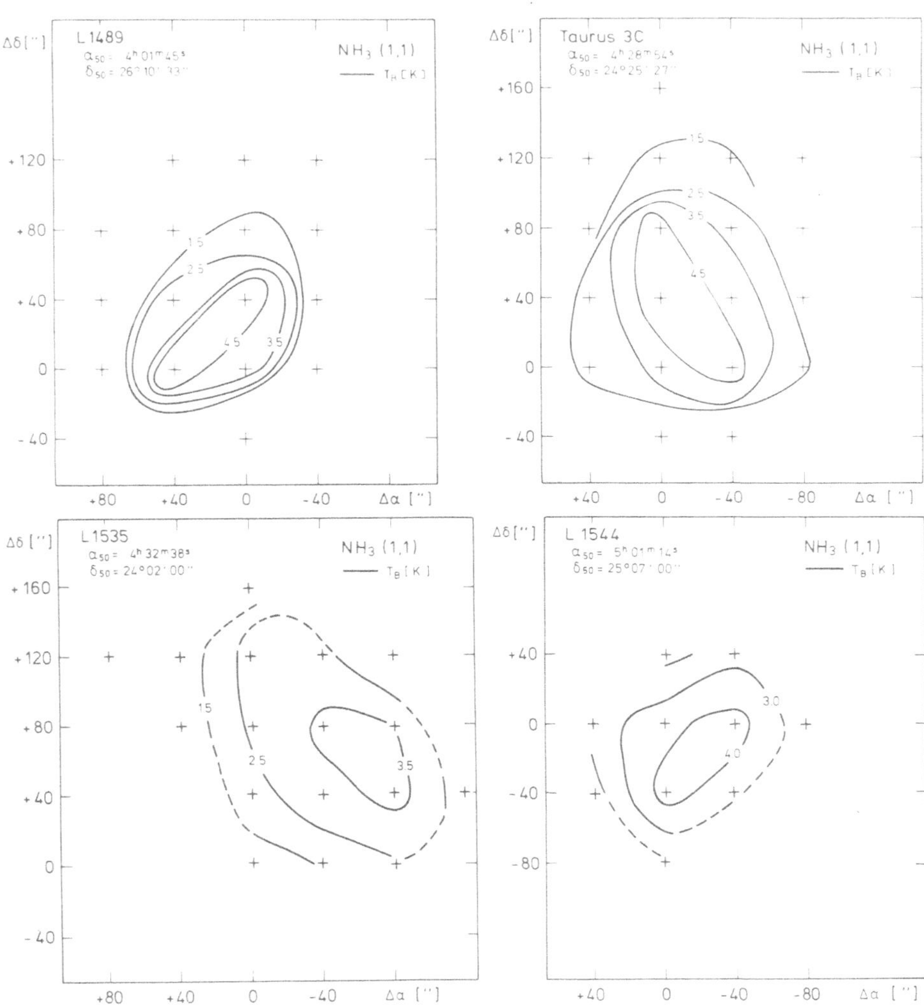

Figure 1

Maps of the peak brightness temperature of the ammonia (J,K)=(1,1) line in cold dense cores of the Taurus molecular cloud complex. (Observations with the Effelsberg 100m telescope, Ungerechts et al., 1982, Gaida et al., 1984).

Physical characteristics of cloud cores derived from NH_3 observations:

Source	Effelsberg, 100m				Haystack, 37m, NRAO, 43m			
	T_{kin}	Δv_{int}	1*d	log n^*	T_{kin}	Δv	R	log n
L1489	9.6	0.22	0.05*0.05	4.8	10	0.28	0.07	4.3
Taurus 3C	8.5	0.24	0.08*0.09	4.8		0.32	0.08	4.7
TMC2	<12	0.24		4.0	10	0.27	0.14	4.4
TMC1	9.5	0.34	0.06*0.60	4.4		0.38	0.23	4.1
L1517	<12	0.18	0.05*0.07	4.2	10	0.20	0.09	3.8
L1544	10	0.34	0.06*0.08	4.1	10	0.32	0.10	4.0
	K	km/s	pc*pc	cm^{-3}	K	km/s	pc	cm^{-3}

References: Effelsberg data : Gaida et al., 1984
Haystack, NRAO : Myers and Benson, 1983

This comparison shows a remarkably close agreement, any differences are smaller than 30% and easily explained if the cloud cores are slightly underresolved with the smaller telescopes: the Effelsberg results tend to give smaller diameters, smaller linewidths and higher densities.

In summary, the cold cloud cores in Taurus have temperatures close to 10K, densities of about $10^{4.5} cm^{-3}$, linewidths 0.2-0.35km/s, and linear sizes 0.05-0.1pc. These paramters imply that they are close to thermal - gravitational equilibrium.

References:

Bastien,P.: 1983, Astron.Astrophys. 119, 109
Benson,P.J., and Myers,P.C.: 1980, Astrophys.J. 242, L87
Benson,P.J., and Myers,P.C.: 1983, Astrophys.J. 270, 589
Elias,J.H.: 1978, Astrophys.J. 224, 857
Gaida,M., Ungerechts,H., Winnewisser,G.: 1984, Astron.Astrophys. 137, 17
Miller,M.:1984, doctoral thesis, University of Cologne
Myers,P.C., and Benson,P.J.: 1983, Astrophys.J. 266 , 309
Ostriker,J.: 1964, Astrophys.J. 140, 1056
Ungerechts,H., Walmsley,C.M., Winnewisser,G.: 1982, Astron.Astrophys. 111, 339

Frosts in the ρ Oph Molecular Cloud

by H Zinnecker[1], A S Webster[1] and T R Geballe[2]

[1] Royal Observatory, Blackford Hill, Edinburgh, Scotland EH9 3HJ

[2] United Kingdom Infrared Telescope Unit, 900 Leilani Street, Hilo, Hawaii 96720.

The existence or otherwise of water ice in the ρ Ophuichi Molecular Cloud presents a puzzle. None has been found towards stars with $A_v < 12.5$ mag (Whittet & Blades 1980) and, although a detection has been claimed in more heavily-obscured objects (Harris et al. 1978), it has been disputed on technical grounds (Whittet & Blades 1980). The matter is of interest because ice can be found in other clouds at much lower values of A_v and because the establishment of a relationship between ice and extinction would help fix the broad-band spectra and luminosities of the embedded stars; this is of great importance because the stars may constitute the only known case of an open cluster in the process of formation (Wilking & Lada 1983).

Accordingly, in 1984 July the two brightest embedded sources (Elias 29 and WL16) were scanned with the Cooled Grating Array Spectrometer on the 3.8m United Kingdom Infrared Telescope in Hawaii. Both were found to show extremely deep absorption in the 3.1 micron band of water ice, with peak optical depths of 1.7 and 1.1 respectively. This finding resolves the uncertainty about whether ice exists in the ρ Oph cloud: it definitely does if one observes objects with sufficiently high extinction ($A_v \sim 47$ and 31 respectively; Lada & Wilking 1984). We hope to extend this work to many more of the embedded stars in order to understand the physics and chemistry of ice-mantle deposition and to improve our knowledge of the stars themselves.

Encouraged by the unusual strength of the ice feature we next looked for the 4.6 micron transition of solid CO and found it in both sources, with peak optical depths of 0.21 and 0.15. This indicates that a substantial part of the dust is extremely cold: Léger (1983) calculates an upper limit of 17K for the grain temperature if CO frost is to form. This low temperature underscores the remarkable quiescence of the cloud indicated previously by the narrowness of the millimetre transitions of the gas, and points to the absence of energetic objects such as massive stars and protostars. The location of the cold dust is unknown: it may be circumstellar, but finding CO absorption in both objects encourages the speculation that it is distributed throughout the core of the cloud. Again, more observations are planned in order to decide this issue. The column density of solid CO is ~ 3-7% of that of gaseous CO as determined by millimetre-wave observations. This relatively small fraction is typical of most sources known to show solid CO (Lacy et al. 1984) but is not easy to understand because the behaviour of the vapour pressure is such that above 17K CO should be almost entirely gaseous and below 17K almost entirely solid (Léger 1983). Léger suggests that a widespread but unknown degassing agency is operating to return most of the frost to the gas phase; we suggest as an alternative possibility that the carrying capacity of the grains is limited in some way and that an arbitrarily-thick mantle of CO frost cannot form.

In some objects showing solid CO absorption Lacy et al. (1984) also found another absorption feature near 4.6 microns which, on the basis of laboratory experiments, was attributed to the stretching vibration of cyano groups in molecules formed by the ultraviolet photolysis of a mixed frost of CO and NH_3. This feature is absent in the two sources we observed, leading us to infer that the cold grains in the ρ Oph cloud have not suffered exposure to a strong ultraviolet flux.

We also found strong Brackett α and weaker Pfund β emission from both objects. The quiescence of the gas makes unlikely the presence of very luminous stars, so we suppose that these recombination lines do not arise in compact HII regions but in the atmospheres of pre-main sequence stars similar in kind to T Tauri.

<u>References</u>

Harris, D.H., Woolf, N.J. and Rieke, G.H., 1978. Astrophys.J. <u>226</u>, 829.

Lacy, J.H., Baas, F., Allamandola, L.J., Persson, S.E., McGregor, P.J., Lonsdale, C.J., Geballe, T.R. and van de Bult, C.E.P., 1984. Astrophys.J., <u>276</u>, 533.

Lada, C.J. and Wilking B.A., 1984. Preprint.

Léger, A., 1983. Astron.Astrophys. <u>123</u>, 271.

Whittet, D.C.B. and Blades, J.C., 1980. Mon.Not.R.astr.Soc. <u>191</u>, 309.

Wilking, B.A. and Lada, C.J., 1983. Astrophys.J. <u>274</u>, 698.

Observations of ^{12}CO Data and Models for the Dark Cloud L183.

K.VEDI[1], I.P.WILLIAMS[1], L.AVERY [2], G.J.WHITE [3], N.CRONIN [4].
[1]School of Mathematical Sciences, Queen Mary College.
Mile End Road, London E1 4NS, England.
[2]National Research Council, Ottawa, Canada K1A 0R6.
[3]Dept of Physics, Queen Mary College, Mile End Road,
London E1 4NS, England.
[4]Dept of Physics, University of Bath, Claverton Down, Bath
BA2 7AY, England.

Abstract.

We present here the first observations of the ^{12}CO $J=3\rightarrow2$ and new $J=2\rightarrow1$ transitions for the dark cloud L183. Theoretical modelling was carried out to fit the observed antenna temperatures and some of the results obtained are included.

Introduction.

The dark cloud L183 ($\alpha = 15^h\ 51^m\ 30^s$, $\delta = -02^\circ\ 43'\ 31''$,1950.0) lies about 2° N and 0.5° W of L134 and is also known as L134N. To date, various molecular line studies of this cloud have been carried out and some of the results will be used here. From the H_2CO measurements of Clark and Johnson (1981), a central density of 10^3-$10^4 cm^{-3}$ is indicated. The angular diameter is 31′ and the core size is estimated to be about 10′ . Assuming a distance of 100pc (Ungerechts, Walmsley and Winnewisser, 1979), the cloud radius is about 0.45pc and a mass of between about 70 and $200 M_\odot$ is estimated. A velocity gradient of about 2±1 km s^{-1} pc^{-1} is inferred and the maximum linewidth found by Caldwell (1979) is 2.5 km s^{-1}. Values for the kinetic temperature within the cloud vary from 9K as determined by the Ammonia measurements of Ungerechts *et al* to 12K from CO line studies by Snell(1981).

Table 1 summarizes known ^{12}CO observations for the $J=1\rightarrow0$ and $J=2\rightarrow1$ transitions and will be used for comparison to the predicted temperatures of our theoretical models.

The Observations.

The ^{12}CO J = 3→2 observations were carried out at the 3.8 m United Kingdom Infra Red telescope (UKIRT) in May 1983, using a QMC millimetre / submillimetre spectral line receiver. The system noise temperature was typically 300 K at 230 GHZ and 500 K at 345 GHZ. Our observing position was $\alpha = 15^h\ 51^m\ 18^s$ and δ = -03° 00′ 00″. The J = 2→1 data was first obtained in October 1983 and again in June 1984 at the 12 m Kitt Peak telescope. The results for our central position are shown in Figures 1 and 2 and further results will be published shortly. There is no evidence from our data to support pedestal features as seen by Frerking and Langer (1982); however the J = 3→2 data was noisy and of a poor quality and further observations are required.

Table 1.

Date of observation	Transition J =	Peak Radiation Temp T_R	Observer	Instrument
1973	1→0	7.0K	Dickman (1975)	5m MWO Ft Davis, Texas
1974-1977	1→0	7.0K	Phillips et al	KPNRAO 11m
	2→1	6-7.0K	" " " (1979)	"
1977	1→0	7-8.0K	Caldwell (1979)	5m MWO Texas
1977	1→0	8.3K	McCutcheon (1980)	4.6m Aerospace Corporation.
1976-1978	1→0	10.0K	Snell (1981)	5m MWO Texas
1983-1984	2→1	6-7.0K	White et al	3.8m UKIRT
	3→2	2-2.5K	" " " (This work)	" "

Figure 1: The $J = 3 \rightarrow 2$ map of radiation temperature against velocity.

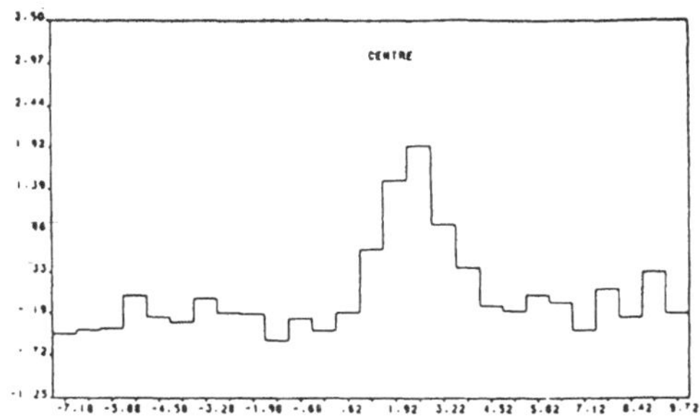

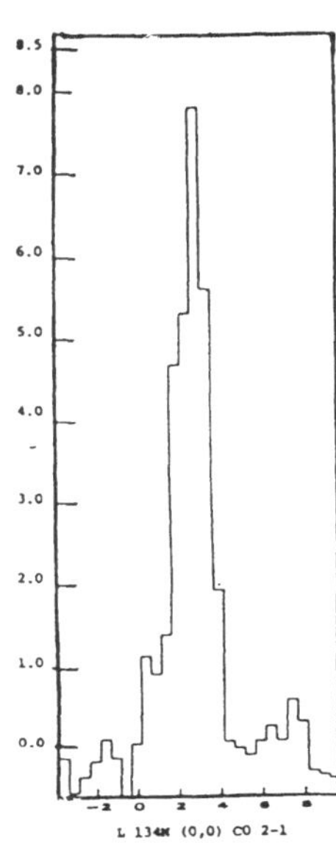

Figure 2: The $J = 2 \rightarrow 1$ map of temp vs velocity

Theoretical Modelling.

We have used Avery's (1983) Monte Carlo radiative transfer programs, based on Bernes's (1979) model, which combine velocity gradients and microturbulence. The model deals with variations of the following parameters throughout the cloud –: kinetic temperature, velocity, molecular abundance, turbulence and hydrogen density and it is possible to set the core density and temperature.

Several models were attempted but matching the radiation temperatures for all three transitions simultaneously proved to be problematical. However, working on two fundamentally different models where:

a) L183 is considered to be a single large cloud, with our line of sight closer to the edge than the centre, and where a radially increasing temperature is attributed to UV heating by the interstellar medium or,

b) the cloud is considered to be a small cold dark clump, within a larger fragmentary complex namely L183, and is isothermal.

After several runs using variations for all the parameters, reasonable fits to the observations were obtained. The best two obtained so far being:

a) A static model where the temperature and density ($\propto T^{-1}$) have values of 5K and 1.5×10^4 cm^{-3} in the core and vary radially to 12K and 6.55×10^3 cm^{-3} in the outer layers. A turbulent velocity of 0.5 Km s^{-1} is incorporated and the CO abundance used is 3×10^{-5}. The predicted radiation temperatures for the three transitions are:

T_R $(1 \rightarrow 0)$ = 7.5 K, T_R $(2 \rightarrow 1)$ = 6.1 K and T_R $(3 \rightarrow 2)$ = 4.1 K, all with linewidths of about 1 Km s^{-1}.

b) An isothermal model with a kinetic temperature of 15 K which is 20% of the size of L183 was formulated. The density (α r^{-3}) varies from 4.6 × 10^4 cm^{-3} in the core to 1.4 × 10^3 cm^{-3} at the edge. It is collapsing with an infall velocity of 0.55 Km s^{-1} and the CO abundance is 2 × 10^{-5}. There is a constant turbulent velocity of 0.3 Km s^{-1}. The predicted radiation temperatures for the three transitions are: T_R (1→0) = 7.9 K, T_R (2→1) = 5.9 K and T_R (3→2) = 3.25 K. All the linewidths are of order 1 Km s^{-1} in accord with the observations.

Discussion.

Both models give a reasonable fit to the data; The first model has a cold core with hot outer layers and is a physically feasible model. The second model is easy to account for by the fragmentary hypothesis and assuming that the lines are produced from one of the denser clumps. However, the problem is that a source may be required to explain the high kinetic temperature as UV heating by the interstellar medium may not be a tenable idea to produce such a temperature (without gradients) throughout the cloud.

This work is still in progress and it is probable that a better correspondence between theoretical and observational results will be obtained.

References.

Avery, L., 1983 Private communication.

Bernes, C., 1979, *Astron.Astrophys.*, 73, 67.

Caldwell, J.A.R., 1979, *Astron.Astrophys.*, 71, 255.

Clark, F.O., and Johnson, D.R. 1981, *Astrophys. J.*, 247, 104.

Dickman, R.L., 1975, *Astrophys.J.*, 202, 50.

Frerking, M.A., and Langer, W.D., 1982, *Astrophys.J.*, 256, 523.

McCutcheon, W.H., Dickman, R.L., Shuter, W.L.H., and Roger, R.S. 1980, *Astrophys. J.*, 237, 9.

Phillips, T.G., Huggins, P.J., Wannier, P.G.,and Scoville, N.Z., 1979, *Astrophys.J.*, 231, 720.

Snell, R.L., 1981, *Astrophys.J.Suppl.Ser.*, 45, 121.

Ungerechts, H., Walmsley, C.M., WinnewiSser, G. 1980, *Astron.Astrophys.*, 88, 259.

PART III

STAR FORMATION IN NEARBY MOLECULAR CLOUDS

DENSE CORES AND STAR FORMATION IN NEARBY DARK CLOUDS

P.C. Myers
Harvard-Smithsonian Center for Astrophysics
60 Garden Street, Cambridge, Massachusetts 02138 USA

ABSTRACT

We review recent progress in the study of dense cores in dark clouds, as revealed by gas emission line observations from NH_3 (1.3 cm), CO (2.6 mm); and by dust and stellar continuum observations with the IRAS satellite (12-100 μm), and with ground-based telescopes (1-5 μm). The main results are: (1) Dense cores are actively forming low-mass stars, typically on time scales ~ 10^5 yr. (2) Stars in cores frequently have CO outflow, which may typically last ~ 10^5 yr; enhanced NH_3 line broadening, of order 0.2 km s^{-1}; and very cold infrared spectra, peaking at $\lambda \gtrsim 100$ μm. (3) Cores without stars have velocity dispersions that are nearly thermal, suggesting that their nonthermal support has largely dissipated.

I. INTRODUCTION

Visually opaque condensations, both isolated and in complexes, have been known since the time of Barnard [1] and were first proposed to form stars by Bok and Reilly [2] over 30 years ago. In the last decade molecular line studies have established the size, temperature, density, velocity dispersion, and mass of such objects within a few hundred pc of the Sun [3-6]. Table 1 summarizes this information, based on observations of emission in the (J,K)=(1,1) line of NH_3 at 1.3 cm [6].

Table 1
Properties of Dense Cores

FWHM Size (pc)	Kinetic Temperature (K)	Number Density (cm^{-3})	FWHM Line Width (km s^{-1})	Mass ($M_\odot$)	Free-fall Time (yr)
0.1	10	$3x10^4$	0.3	1	$2x10^5$

In this report we describe recent work, which suggests that such cores are actively forming low-mass stars, probably on the time scale ~ 10^5 yr. Cores can now be divided into those known to have associated stars and those known not to have associated stars. We discuss distinguishing properties of each group, emphasizing information in molecular line widths.

II. CORES FORM STARS

The evidence that cores with properties in Table 1 form low-mass stars has become stronger as core positions and sizes have been compared with stellar data at optical, near-infrared, and far-infrared wavelengths. In Taurus-Auriga, T Tauri stars were found to be clustered in loose groupings 1-3 pc in size [7,8], and each of the seven groups studied was found to contain one or two cores in projection [6]. At 2 μm, a survey of 25 cores revealed six to have an associated star within one core map diameter of the map peak [9]. Four of these six stars are optically invisible due to obscuration. At 12 μm, IRAS results indicate that 12 of 23 cores have a star within one core map diameter of the map peak [10,11]. Most of these 12 stars were not detected in the original 2 μm searches [9] but have since been detected in more sensitive 2 μm searches over smaller areas [12]. Most of these 12 stars were also detected by IRAS at 25, 60, and 100 μm and have very similar spectra. Furthermore, many of them have evidence of interaction with the core gas in the molecular line widths and maps, as described in Section III. Thus, there is little chance that they are field stars, unrelated to the cores on which they are projected.

These detection statistics allow a crude estimate of the "waiting time" that elapses before a typical star-free core begins its star-forming collapse. Suppose that the evolution of a core can be described by the four times t_o, when the gas becomes dense enough to be detectable by the NH_3 surveys of [6]; t_1, when the star-forming collapse begins; t_2, when the star becomes visible to IRAS at 12 μm; and t_3, when the star becomes visible optically, e.g., visible on photographs having the sensitivity of the Palomar Sky Survey. Let us call $t_1 - t_o \equiv \tau_{wait}$, $t_2 - t_1 \equiv \tau_{collapse}$, and $t_3 - t_2 \equiv \tau_{clear}$. Then the likelihood of finding an optically invisible 12 μm IRAS source in a core is

$$p = \frac{\tau_{clear}}{\tau_{wait} + \tau_{collapse} + \tau_{clear}} , \qquad (1)$$

whence

$$\tau_{wait} = \tau_{clear}(p^{-1} - 1) - \tau_{collapse} . \qquad (2)$$

It is not possible to estimate either τ_{clear} or $\tau_{collapse}$ in eq. (2) with much precision, but it is likely that each of them is of order 10^5 yr, or less. The time $\tau_{collapse}$ should be less than, or of the order of, the free-fall time, typically 2×10^5 yr for dense cores. The time $\tau_{collapse} + \tau_{clear}$ should be the age of the youngest T Tauri stars, again $\sim 2 \times 10^5$ yr according to the observations and models in [7], and according to the "birthline" for low-mass stars described in [13]. Since $p \sim 0.5$ from the detection statistics reported here, it follows that $\tau_{wait} \lesssim 10^5$ yr. Thus in this model the waiting time is comparable, or less than, the free-fall time.

This result implies that cores without stars generally are now forming, or will soon form, stars: there is no inactive period long compared to the free-fall time. Consequently cores without stars are good candidates for observations seeking evidence of infall or other motions associated with very early stellar evolution.

These results do not bear on the closely related questions of how, or how long, cores evolve before they are dense enough to be detected in surveys of emission lines such as the NH_3 line, requiring gas density $\gtrsim 10^4$ cm^{-3}. This period could conceivably be much longer than 10^5 yr.

III. PROPERTIES OF CORES WITH STARS

Cores with stars show at least two types of interaction between the stars and their cores. In addition, stars in cores have remarkably cold infrared spectra. In this section we describe recent observations that demonstrate these results.

A. CO Outflows

The association of a young star with spatially displaced maps of CO emission in high- and low-velocity line wings is now a well-known signature of stellar outflow, a very early phase of stellar evolution. CO surveys for evidence of outflow have detected 3 outflows in 180 opaque condensations whose stellar content was unknown [14; detection rate 2%] and 3 outflows in 28 T Tauri stars whose degree of association with dense cores was unknown [15; 11%]. Recently a survey of dense cores known to have young associated stars gave a significantly higher detection rate: of eight such cores, four have CO outflows [16]. These outflows are similar in spatial and velocity extent to others seen in dark clouds [17]. They are described in Table 2.

Table 2
Newly Found Outflows from Dense Cores with Embedded Stars [16]

Core	Stellar Position α(1950)	Stellar Position δ(1950)	CO LSR Velocity (km s^{-1})	Velocity Extent (km s^{-1})	Adopted Distance (pc)	Spatial Extent (pc)
L1489	04 01 40.6	26 10 48.9	6.0	6	140	0.3
L43	16 31 37.9	-15 40 50.0	0.0	10	160	0.4
L778	19 24 26.4	23 52 37.0	10.0	19	200	0.5
L1172	21 01 48.6	67 42 13.0	3.0	15	440	0.9

If one adds to these cases dense cores with embedded stars that have previously been searched for CO outflow (L1551, L1455, B335), then the fraction of such cores with outflows is 7/11. This is a remarkably high fraction, since in any sample some outflows are probably not detected due to the orientation of the outflow direction in the plane of the sky or along the line of sight. This comparison of detection rates suggests that stars in cores are probably younger than T Tauri stars, a conclusion consistent with the idea that young stars clear away their parent cores, perhaps via outflows, before becoming visible as T Tauri stars. If so, the high detection rate of outflows from stars in cores suggests that the outflow process typically lasts for a significant fraction of the T Tauri age -- a time closer to 10^5 yr than to the $\sim 10^4$ yr dynamical time deduced from outflow velocity and spatial extent.

B. NH_3 Line Broadening

The presence or absence of embedded stars is now known for some ~ 25 cores, and it is therefore possible to compare the gas properties of the two groups for differences that might reflect the presence or absence of star-core interaction. We have compared the "turbulent" part of the NH_3 line broadening, defined by

$$\Delta v(\mathrm{turb}) \equiv \left[\Delta v(\mathrm{obs})^2 - 8 \ln 2\, kT/m\right]^{\frac{1}{2}} , \qquad (3)$$

where Δv(turb) and Δv(obs) are the FWHM turbulent and observed widths, T is the gas kinetic temperature, and m is the NH_3 molecule mass, 17 amu. For sensitive NH_3 line measurements, this quantity is known with a relative precision of $\lesssim$ 10% [6,18].

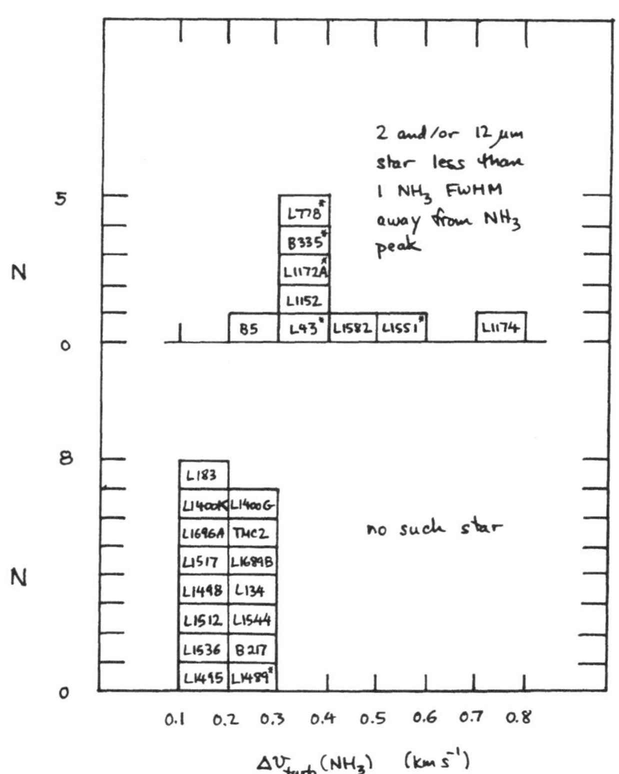

Figure 1. Distributions of the turbulent component of the FWHM NH_3 line width, defined in eq. (3), for cores with embedded stars (upper) and without embedded stars (lower). Core names with asterisks have evidence of CO outflow.

Figure 1 is a histogram of values of Δv(turb) for cores with and without stars. The two groups are distinctly different, with Δv(turb) ~ 0.4 km s^{-1} for cores with stars and ~ 0.2 km s^{-1} for cores without stars. This difference is also evident in other lines, including $C^{18}O$ and ^{13}CO [19], although in these lines the velocity width difference is of greater magnitude and is less distinct than in the NH_3 line.

The origin of the difference in NH_3 line widths in Figure 1 is unclear, but it appears likely that the greater widths reflect the interaction of the associated stars with their cores. The mechanical energy associated with a typical core having Δv(turb) = 0.4 km s^{-1} exceeds that of a core having Δv(turb) = 0.2 km s^{-1} by ~7 x 10^{41} erg, or less than one percent of the typical energy of a molecular outflow in dark clouds [17]. Since many of the cores with stars have outflows, as noted in Section III A, it seems plausible that in some cases a small fraction of the outflow energy couples into the dense gas seen in the NH_3 line and increases its turbulence. Another possible mode of star-core interaction that could account for the observed difference in Δv(turb) is gravitation. At the typical dense core radius r = 0.05 pc, the free-fall radial speed and the circular orbit speed for a central mass $M = 1M_{\odot}$ are each of order $\sqrt{GM/r}$, or 0.2 km s^{-1}. To resolve these and other questions, it will be necessary to study the cores with finer angular resolution, $\lesssim$ 10", than has previously been available.

C. Infrared Spectra

The 2–100 μm spectra of young stars in or near dense cores have remarkably similar shapes, with flux density increasing with wavelength λ as $\lambda^{1.5-2.0}$. For a given star, the 12–100 μm IRAS spectra [11] and 1–5 μm near infrared spectra [12] are usually consistent with each other in amplitude and slope. The 2–100 μm luminosity is generally $\lesssim$ 10 $L_\odot$, and the spectra probably "turn over" at a typical wavelength of 100–300 μm. This suggests a substantial amount of dust with temperature ~ 30 K at radius ~ 10^{16} cm, if the emission is due to a blackbody shell. Preliminary statistics indicate that the sources with the strongest 100 μm emission are centered in their cores and have CO outflows.

IV. PROPERTIES OF CORES WITHOUT STARS

As noted in Section II, cores without stars have values of Δv(turb) significantly smaller than cores with stars, and the stellar detection statistics suggest that cores without stars are now forming, or will soon (within ~ 10^5 yr) form, low-mass stars. Here we show that the typical values of Δv turb) in cores without stars are sufficiently small that the internal motions on the size scale ~ 10^{17} cm must be almost entirely thermal.

For the nine cores listed in Figure 1 with neither a 2 μm nor an IRAS point-source detection, the values of Δv(turb) range from 0.11 km s^{-1}in L1498 and L1517 to 0.29 km s^{-1} in L1681B, with the median value being 0.20 km s^{-1}. For a typical core with kinetic temperature T = 10 K, the gas particle of mean mass μ = 2.33 amu has a distribution of thermal velocities with FWHM $\Delta v(\text{thermal}) = (8 \ln 2\ kT/\mu)^{1/2} = 0.44$ km s^{-1}. Consequently the starless core with median Δv(turb) has nonthermal motions with Mach number $\underline{M} = \Delta v(\text{turb})/\Delta v(\text{thermal}) = 0.45$, or equivalently, its thermal motions represent $(1 + \underline{M}^2)^{-1/2} = 0.90$ of its total internal motions. Similarly each of the cores with minimum Δv(turb), L1498 and L1517, has $\underline{M}$ = 0.25, and thermal motions represent 0.97 of its total internal motions.

For these cores nonthermal motions are clearly unimportant for support. If they evolved from lower-density condensations resembling those studied in lines of ^{13}CO and C^{18}O [19], the nonthermal motions have reduced dramatically while the thermal motions have stayed nearly constant. One may speculate that this reduction in nonthermal motions may be a characteristic, or perhaps necessary, stage in the evolution of a core toward collapse and star formation [18].

If, as suggested above, some starless cores have begun to collapse, are their low values of Δv(turb) consistent with the expected collapse motions? As noted earlier the free-fall speed from 0.05 pc onto 1 $M_\odot$ is 0.2 km s^{-1}, so the values of Δv(turb) are roughly consistent with collapse speeds. However, the separation of Δv(turb) into random and systematic components must await observations of finer angular resolution.

REFERENCES

[1] Barnard, E. 1927, Atlas of Selected Regions of the Milky Way, eds. E. Frost and M. Calvert (Washington: Carnegie Institution).
[2] Bok, B., and Reilly, E. 1947, Small Dark Nebulae. Ap. J., 105, 255–257.
[3] Churchwell, E., Winnewisser, G., and Walmsley, C. 1978, Molecular Observations of a Possible Proto-Solar Nebula in a Dark Cloud in

Taurus. _Astr. Ap._, _67_, 139-147.
[4] Martin, R., and Barrett, A. 1978, Microwave Spectral Lines in Galactic Dust Globules. _Ap. J. (Suppl.)_, _36_, 1-51.
[5] Snell, R. 1981, A Study of Nine Interstellar Dark Clouds. _Ap. J. (Suppl.)_, _45_, 121-175.
[6] Myers, P., and Benson, P. 1983, Dense Cores in Dark Clouds. II. NH_3 Observations and Star Formation. _Ap. J._, _266_, 309-320.
[7] Cohen, M., and Kuhi, L. 1979, Observational Studies of Pre-Main-Sequence Evolution. _Ap. J. (Suppl.)_, _41_, 743-843.
[8] Jones, B., and Herbig, G. 1979, Proper Motions of T Tauri Variables and Other Stars Associated with the Taurus-Auriga Dark Clouds. _A. J._, _84_, 1872-1889.
[9] Benson, P., Myers, P., and Wright, E. 1984, Dense Cores in Dark Clouds: Young Embedded Stars at 2 Micrometers. _Ap. J. (Letters)_, _279_, L27-L30.
[10] Beichman, C. 1984, IRAS Observations of Solar Type Stars. Talk presented at _Protostars and Planets. II._, Tucson.
[11] Beichman, C., _et al._ 1985, in preparation.
[12] Myers, P., Benson, P., Mathieu, R., Fuller, G., Fazio, G., and Beichman, C. 1985, in preparation.
[13] Stahler, S. 1983, The Birthline for Low-Mass Stars. _Ap. J._, _274_, 822-829.
[14] Frerking, M., and Langer, W. 1982, Detection of Pedestal Features in Dark Clouds: Evidence for Formation of Low-Mass Stars. _Ap. J._, _256_, 523-529.
[15] Edwards, S., and Snell, R. 1982, A Search for High-Velocity Molecular Gas Around T Tauri Stars. _Ap. J._, _261_, 151-160.
[16] Myers, P., Hemeon-Heyer, M., Snell, R., and Goldsmith, P. 1985, in preparation.
[17] Goldsmith, P., Snell, R., Hemeon-Heyer, M., and Langer, W. 1984, Bipolar Outflows in Dark Clouds. _Ap. J._, _286_, 599-608.
[18] Myers, P. 1983, Dense Cores in Dark Clouds. III. Subsonic Turbulence. _Ap. J._, _270_, 105-118.
[19] Myers, P., Linke, R., and Benson, P. 1983, Dense Cores in Dark Clouds. I. CO Observations and Column Densities of High-Extinction Regions. _Ap. J._, _264_, 517-537.

IRAS OBSERVATIONS OF STAR FORMATION IN NEARBY MOLECULAR CLOUDS

C. A. Beichman (Jet Propulsion Laboratory), J. P. Emerson (Queen Mary College),

R. E. Jennings (University College London), S. Harris (Queen Mary College),

B. Baud (Kapteyn Institute) and E. T. Young (University of Arizona)

Introduction

Data from IRAS[1] have been used to investigate the problem of the formation of low-mass stars in nearby molecular clouds. With wavelength coverage at 12, 25, 60 and 100 μm and 1σ sensitivities of 0.1, 0.1, 0.2 and 0.5 Jy, the IRAS survey is sensitive to emission from point sources emitting as little as 0.1 $L_{\odot}$ in the temperature range 50-300 K at typical distances of 150 pc. Pointed observations, which offer a three- to five-fold increase in sensitivity and modest increases in spatial resolution, were used extensively in this investigation. A fuller discussion of these data is in preparation (Beichman *et al.* 1985).

The Taurus and Ophiuchus molecular clouds were chosen for particular attention because they lie well above the galactic plane, thus reducing confusion with background emission; because they are close, 140 and 160 pc respectively, which results in excellent spatial resolution (1' = 0.05 pc) and maximum sensitivity to intrinsically faint objects; and because they have been well studied in the millimeter spectral lines of a large number of molecular species so that much is known about the gas content of these clouds.

Despite their similarities, the Taurus and Ophiuchus regions have striking differences as shown in the first two figures. On the largest scale (30°) Taurus (Fig. 1) appears to be very diffuse and filamentary (cf. Herbig 1977), while Ophiuchus (Fig. 2) appears much more centrally condensed, with a bright core that is known, from previous earth-bound observations, to be the site of the formation of intermediate mass stars (cf. Wilking and Lada 1984). Young stars abound in the two regions as demonstrated by the optical studies which reveal the presence of T Tauri stars (e.g. Cohen and Kuhi 1979) and Herbig Haro objects and far infrared studies which reveal the presence of low luminosity stars embedded within the molecular clouds (Fazio *et al.* 1976; Cudlip *et al.* this conference).

What types of star form in small, nearby molecular clouds and what are the physical conditions that lead a cloud to produce a star? To answer these questions in a systematic way we have examined data from IRAS pointed observations made toward more than 90 molecular cloud cores observed in CO and NH_3 by Myers and his collaborators (Myers, Linke and Benson 1983,

[1]The ***Infrared Astronomical Satellite*** was developed and operated by the Netherlands Agency for Aerospace Programs (NIVR), the US National Aeronautics and Space Administration (NASA) and the UK Science and Engineering Research Council (SERC). Some of this work was performed for NASA by the Jet Propulsion Laboratory of the California Institute of Technology.

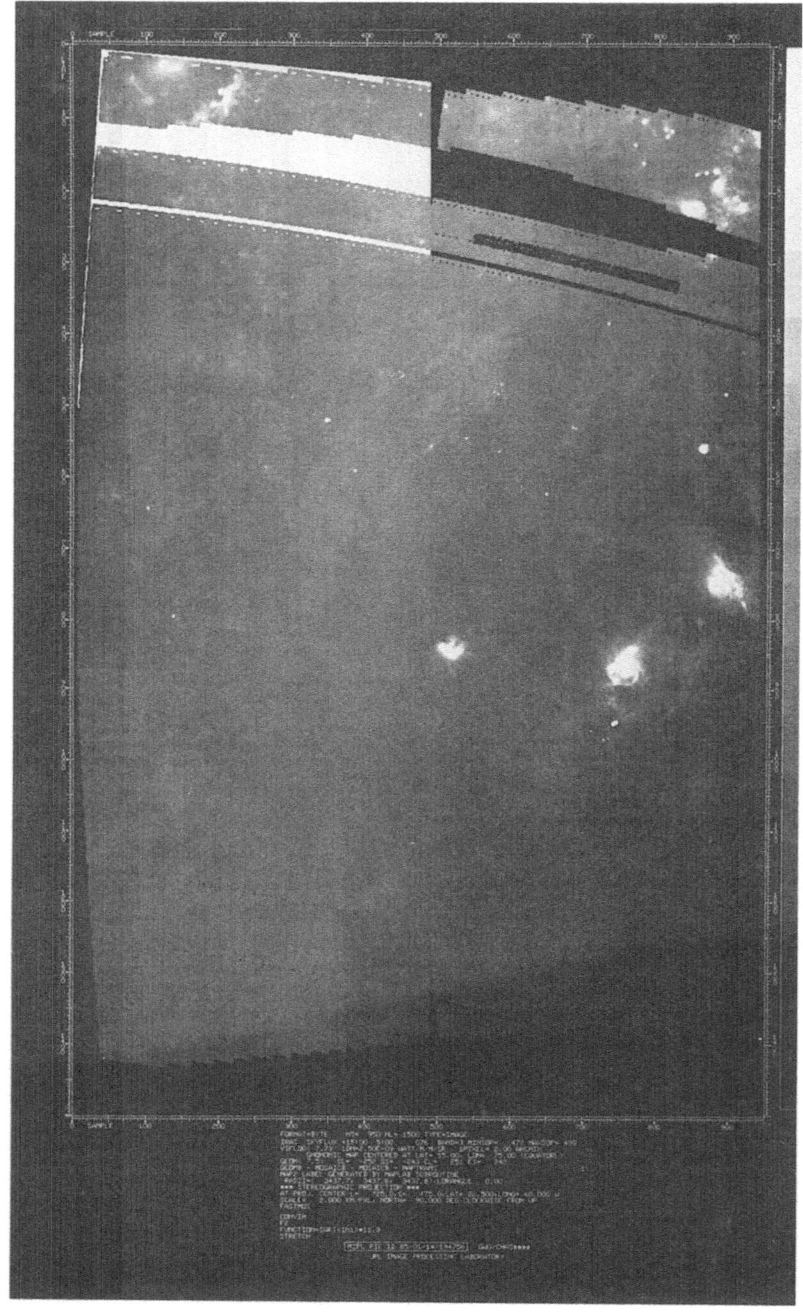

Figure 1. A large scale view of the Taurus-Perseus region showing a region approximately 45° by 30° in size centered on $\alpha=5^h$ and $\delta=15°$ with North up and East to the left. The image shows the 60 μm emission from the region. Note the diffuse, filamentary structure. The bright nebula in the center is the Pleiades. The California Nebula and IC 348 are at the top and the λ Ori HII region is visible at the lower left. Many of bright points in the left hand half of the picture are the point sources of infrared emission associated with the molecular cloud cores discussed in the text.

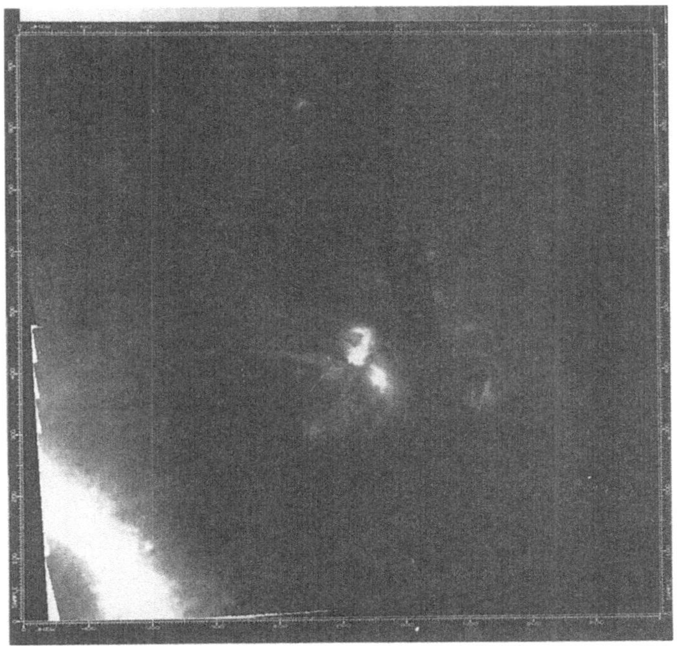

Figure 2. A large scale view of the Ophiuchus-Sco-Cen region showing an approximately 30° square centered on $\alpha=17^h$ and $\delta=-30°$ with North up and East to the left. The image shows the 60 μm emission from the region. Note the centrally concentrated nature of the ρ Oph cloud. The Galactic plane is visible at the lower left.

hereafter MLB; Myers and Benson 1984, hereafter MB). This sample, chosen for having a compact, opaque appearance on the Palomar Observatory Sky Survey prints (POSS) was found from CO observations to have a mean size of 0.3 pc, a mass of 35 $M_\odot$ and a gas kinetic temperature of 15 K. Often contained within the CO core was an even denser core, detected in NH_3, that was smaller and less massive, containing only a few $M_\odot$. The millimeter line data suggested that these clouds were in approximate virial equilibrium.

IRAS Results

The result of comparing the 90 molecular cloud cores contained in the lists of MLB and MB with the IRAS maps was the discovery of a large number of point-like sources of infrared emission (i.e. with dimensions less than roughly 0.75' at 12 and 25 μm, less than 1.5' at 60 μm and 3' at 100 μm). A search was made for all sources within 6' (typically 0.3 pc) of the position of the cloud core. As defined by a combination of their energy distribution and the presence of optical counterparts on the POSS, it is possible to identify a few broad categories of objects, similar to the groupings suggested by Beichman *et al.* (1984) for the Barnard 5 cloud and by Baud *et al.* (1984) for the Chamaeleon region. These groupings are summarized in Table 1 and are discussed below. The table lists the number of sources of each type, its defining characteristics, the average color temperatures and luminosities between 12 and 100 μm. The uncertainties listed in the table are 1σ population dispersions.

Table 1. IRAS Observations of Molecular Cloud Cores

Source Type	Number of Sources	Wavelengths Detected (μm)	T(12-25 μm) (K)	T(25-60 μm) (K)	T(60-100 μm) (K)	L(IRAS) ($L_\odot$)
BLANK[1]	24	12-100 25-100	211±13	88±4	43±2	5.5±1.6
REFL[2]	7	12-100	268±10	106±12	42±5	14±11[3]
PMS[4]	16	12-60 12-25	239±6	120±7	---	1.0±0.3
STAR[5]	14	12 and 25 12 or 25	600±85	---	---	0.11±0.06
WARM[6]	7	60-100	---	---	26±2	0.7±0.2
CIRRH	15	60	---	---	---	0.30±0.16
CIRRC	41	100	---	---	---	0.17±0.03

[1] $f_\nu(100\mu m)>f_\nu(60\mu m)>f_\nu(25\mu m)>f_\nu(12\mu m)$; no visible star or nebulosity.
[2] $f_\nu(100\mu m)>f_\nu(60\mu m)>f_\nu(25\mu m)>f_\nu(12\mu m)$; visible reflection nebulosity.
[3] Excluding L1450B, average luminosity is 2.2±1.2 $L_\odot$
[4] $f_\nu(25\mu m)>f_\nu(12\mu m)$ and $f_\nu(25\mu m)\geq f_\nu(60\mu m)$; visible star.
[5] $f_\nu(12\mu m)>f_\nu(25\mu m)$; visible star.
[6] $f_\nu(100\mu m)>f_\nu(60\mu m)$

Embedded sources

These objects are detected at 25, 60 and 100 μm and often at 12 μm and show an energy distribution that rises steeply to longer wavelengths (Fig. 3). The energy distributions show a broad range of color temperatures between the various wavelengths (40-250 K) as expected for

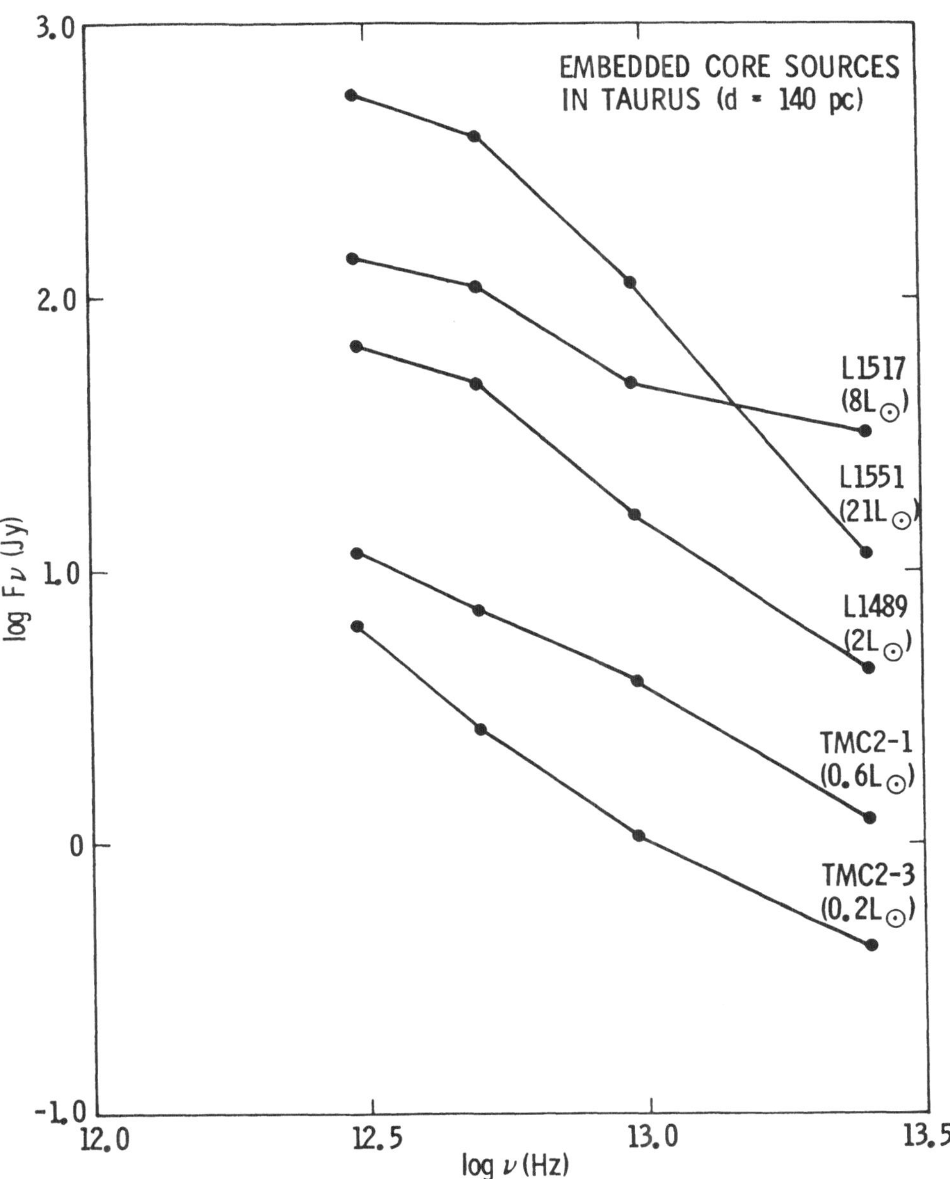

Figure 3. Representative energy distributions from 12 to 100 μm for embedded sources measured by IRAS.

a dust cloud heated by a central energy source. Based on the appearance of the region on the POSS these sources can be divided into those that are (class REFL in Table 1) or are not (class BLANK) associated with visual reflection nebulosity. As shown in Table 1 the 12-25 μm color temperatures of the infrared sources associated with reflection nebulae are considerably higher than those are completely embedded in the surrounding molecular cloud.

Stars

Objects detected only at 12, 25 and occasionally at 60 μm are usually associated with stars visible on the POSS. Stars can be divided into two groups based on the ratio of 12 to 25 μm flux densities. Stars (class PMS) with $f_\nu(25\mu m)>f_\nu(12\mu m)$ are also often detected at 60 μm and have $f_\nu(25\mu m)>f_\nu(60\mu m)$. These are most likely pre-main sequence stars, such as T Tauri stars, which have emerged from their parent clouds but which are often associated with infrared excesses from an optically thin shell or disk of emitting material (Harvey, Thronson and Gatley 1979; Cohen 1983). Obviously, this identification awaits confirmation by optical spectroscopy. Stars detected only at 12 μm or at 12 and 25 μm with $f_\nu(12\mu m)>f_\nu(25\mu m)$ may be field stars or pre-main sequence stars with weaker or hotter infrared excesses (class STAR).

Cirrus

Many sources are detected only at 60 and/or 100 μm. The dust emission from these regions appears on large scale images (Figs. 1 and 2) as having structure on all spatial scales (called "cirrus" by Low *et al.* 1984). Cirrus with structure on the 1'-2' scale may masquerade as true point sources and represent no more than small-scale column density and/or temperature fluctuations. Beichman *et al.* (1984) argue that these are unlikely to be collapsing objects. However, without observations at higher spatial resolution it is impossible to rule out the possibility that some of these might be very cold versions of the embedded sources which were not detected at 12 or 25 μm. Three types of cirrus are observed as shown in Table 1: objects seen at 60 and 100 μm (class WARM) , hot cirrus (CIRRH) seen only at 60 μm and cold cirrus (CIRRC) seen only at 100 μm.

Discussion

In this section the nature of the IRAS sources embedded within the molecular cloud cores is addressed. Since the IRAS sources are usually located within a few tenths of a parsec of the densest part of the surrounding clouds, it is unlikely that they are late-type field stars embedded within a dense region. For typical M and K star densities, < 1 pc^{-3}, the probability of a passing star occupying this privileged position by chance is small.

Although the most likely possibility is that the embedded IRAS sources are newly formed, or forming, low mass stars, their exact evolutionary state is ambiguous. To help understand these objects it is useful to place them in the HR diagram (Fig. 4) using their luminosity between 12 and 100 μm and a color temperature derived from the 12 and 25 μm measurements, or, if 12 μm emission was not detected, from the 25 and 60 μm data. There are drawbacks to this way of estimating both the temperature and luminosity. The very broad energy distributions indicate the presence of steep temperature gradients within these embedded sources. This result is contrary to some theoretical expectations, e.g., Stahler, Shu and Taam (1980), which predict that a single temperature "dust photosphere" would form around accreting protostars. Since no unique temperature exists, the *hottest* color temperature available has been used since that represents the temperature of the material closest to the embedded object.

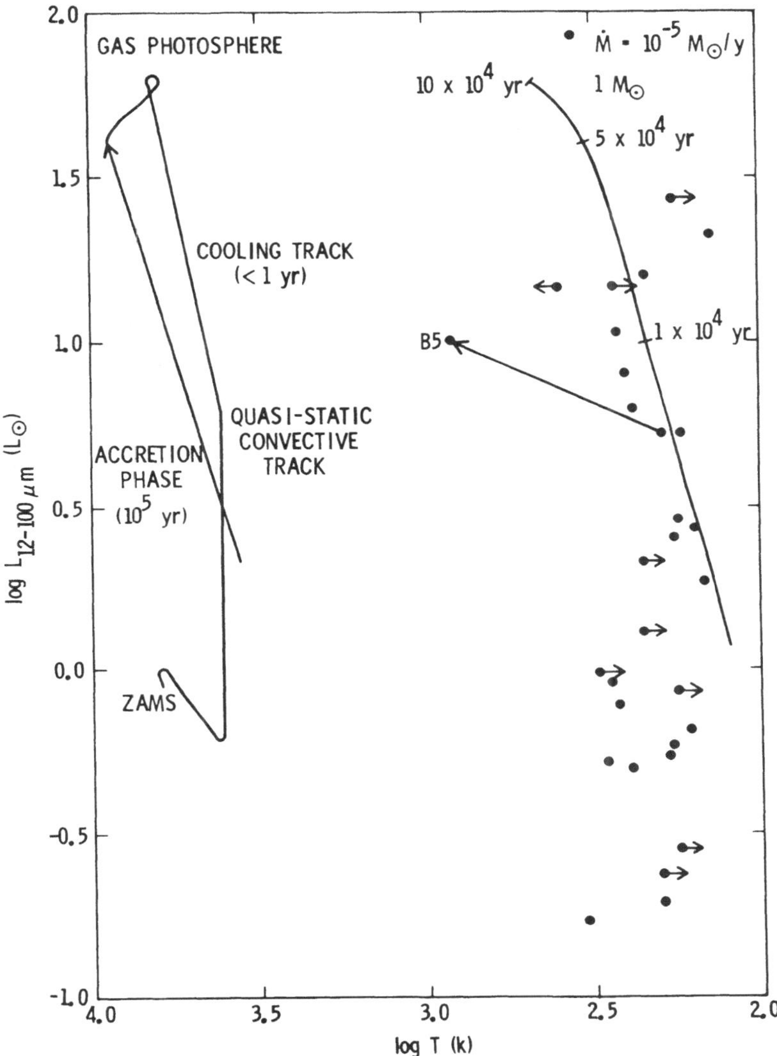

Figure 4. An H-R diagram for IRAS sources found in nearby molecular cloud cores. As described in the text the location of core sources in the plot is obtained from their IRAS properties. For B5-IRS 1, a line connects the location of the source based on its IRAS properties to its location in the plot based on more complete ground-based and IRAS measurements. The line at the left shows the evolution of the photosphere of an accreting proto-star, which is however, invisible. The line at the right shows the properties of the surrounding "dust photosphere". Both theoretical lines are based on the models of Stahler, Shu and Taam (1984).

Another problem is that the 12-100 μm luminosity may not accurately represent the bolometric luminosity of the object, since radiation emitted at wavelengths shorter than the 12 μm and longer than 100 μm is at present unaccounted for. The effect of ignoring this radiation in the case of B5-IRS1 is about a factor of two in luminosity with most of the excess coming at wavelengths between 1.2 and 12 μm (Beichman *et al.* 1984; Benson, Myers and Wright 1984). In the HR diagram B5-IRS1 moves to the upper left (hotter and more luminous) when the effects of emission at non-IRAS wavelengths are included. Effects of a similar magnitude can be expected for most of the brighter sources.

Figure 4 includes a theoretical track (Stahler *et al.* 1980) for a star of $M < 1 M_\odot$ forming out of a cloud with an infall rate of $10^{-5} M_\odot$ yr^{-1}. The track at the left shows the evolution of the invisible photosphere. The luminosity increases steadily as the infall progresses until, after 10^5 yr, the infall ends and the star begins its slow (> 10^5 yr) contraction phase down the convective, Hayashi track to the main sequence. What one can observe during this time is, however, limited by the presence of the parental molecular cloud. The evolutionary track at the right of the figure shows the appearance of the "dust photosphere" that forms about 10^{14} cm away from the star itself. All of the stellar luminosity is absorbed by this dust shell and converted into infrared energy. It is with this second curve that the IRAS observations must be compared.

One cannot determine from the infrared data alone whether these objects are in the early accretion phase or on the descending convective track. The fact that most of these objects more luminous than $1 L_\odot$ are associated with mass outflows as determined from measurements of broadened CO line wings (Myers 1984, this conference) suggests that for these objects the accretion process has ended, perhaps due to the development of a stellar wind; and that these are pre-main sequence stars (e.g. T Tauri stars) still obscured by the surrounding clouds, similar to L1551-IRS 5 (Beichman and Harris 1981).

The nature of the fainter objects, $L < 1 L_\odot$, is less clear. According to standard models, low mass stars approach the main sequence at a constant temperature and steadily decreasing luminosity. The embedded sources could be late-type stars, K or M spectral types with $M < 1 M_\odot$ and $L < 1 L_\odot$ seen relatively late in their evolution down the Hayashi track. A population of young stars with the appropriate range of luminosities, $0.1 - 1.0 L_\odot$, is seen in HR diagrams of objects found visually in the Taurus-Auriga complex (Cohen and Kuhi 1979). One problem with identifying the IRAS sources as younger, still embedded, group of newly-forming K and M stars is the very fact that they are still embedded within the molecular clouds. From the iso-chrones plotted in the Cohen and Kuhi HR diagrams it is clear that stars with this range of luminosity are old enough that it is unlikely that would still be seen close to the sites of their formation. For example, a $0.5 M_\odot$ star takes about 10^6 yr to reach $1 L_\odot$. With relative cloud-star velocities of only a few tenths of a km s^{-1} the star could have escaped the cloud (r=0.3pc) in less than 10^6 yr. For the lowest luminosity IRAS sources, the explanation of a well-formed, but still embedded pre-main sequence star is not entirely satisfactory.

While it is likely that some of these IRAS sources are, indeed, just more deeply embedded versions of the low luminosity stars already found in abundance in Taurus and Ophiuchus complexes, the possibility that some of these objects are deriving their energy solely by accreting material from the molecular cloud must be considered. The luminosities and temperatures of some of the low luminosity IRAS sources are consistent with the gravitational potential energy of material accreting onto a $0.1-0.5 M_\odot$ object at a rate of about $10^{-5}-10^{-6} M_\odot$ yr^{-1}, hitting an accretion shock at a distance from the central object of about 10^{14} cm.

Deciding on the evolutionary state of these low luminosity objects will depend on obtaining additional information at higher spectral and spatial resolution. What is clear, however, is that IRAS has observed many hitherto unobserved embedded young stars of a solar mass and less and may have found stars in the earliest period of formation, during their accretion phase.

Acknowledgements

Too many people contributed to the success of IRAS to be able to acknowledge them all individually. It is appropriate, however, to thank the people who scheduled the additional observations used in obtaining the data described here, in particular K. King and W. Cudlip. We also thank P. Myers and B. Elmegreen for valuable discussions.

References

Baud *et al.* 1984, *Ap. J.*, **278**, L53.

Beichman, C. A. and Harris, S. 1981, *Ap. J.*, **245**, 589.

Beichman, C. A. *et al.* 1984, *Ap. J.*, **278**, L45.

Beichman, C. A. *et al.* 1985 in preparation.

Benson, P. J., Myers, P. C. and Wright, E. L. 1984, *Ap. J.*, **279**, L27.

Chester, T. J. *et al.* 1984, in *The IRAS Explanatory Supplement*, eds. Beichman, C. A., Neugebauer, G., Habing, H. J., Clegg, P. E. and Chester, T. J. (preprint), page V-1.

Cohen, M. and Kuhi, L. V. 1979, *Ap. J. (Suppl)*, **41**, 743.

Cohen, M. 1983, *Ap. J.*, **270**, L69.

Fazio, G. G., Wright, E. L., Zelik, M., III, and Low, F. J. 1976, *Ap. J.*, **206**, L165.

Harvey, P. M., Thronson, H. A. and Gatley, I. 1979, *Ap. J.*, **231**, 115.

Herbig, G. H. 1977, *Ap. J.*, **214**, 747.

Low, F. J. *et al.* 1984, *Ap. J.*, **278**, L19.

Myers, P. C. and Benson, P. J. 1983, *Ap. J.*, **266**, 309.

Myers, P. C., Linke, R. and Benson, P. J. 1983, *Ap.J.*, **264**, 517.

Stahler, S. W. 1983, *Ap. J.*, **274**, 822.

Stahler, S. W., Shu, F. H. and Taam, R. E. 1980, *Ap. J.*, **241**, 637.

Wilking, B. A. and Lada, C. J. 1984, *Ap. J.*, **274**, 698.

Star Formation in the Centrally Condensed Core of the Rho Ophiuchi Dark Cloud

B. A. Wilking
Department of Physics
University of Missouri-St. Louis
8001 Natural Bridge Rd.
St. Louis, MO 63121

I. Introduction

The ρ Ophiuchi dark cloud is one of the nearest (160 pc) and most active star-forming complexes. Devoid of the disrupting effects of massive embedded stars ($M>10M_\odot$), the cloud is ideal for investigations of low mass star formation. A recent wealth of infrared and molecular-line data in the central regions of the ρ Oph cloud have vastly improved our understanding of the molecular structure of the core region and its relationship to the associated star formation. We are now in a powerful position to explore the global aspects of star formation in the ρ Oph cloud and to gain a better understanding of the low mass star formation process in general. Among the key questions which may now be addressed is has enough molecular gas been converted into stars in the ρ Oph cloud core to result in the formation of a gravitationally bound, gas-free cluster? Does the luminosity function of this young cluster deviate significantly from that observed in visible open clusters? In the absence of massive stars, what heating mechanism produces the prevailing warm gas temperatures (T_{gas} ~40-50 K) in the cloud core?

A brief review will be presented of our current state of knowledge concerning the molecular cloud structure and the population of embedded stars in the central core of the ρ Oph dark cloud. The star-forming characteristics of the cloud will be discussed in terms of the formation of a bound cluster of low-luminosity stars. The first attempt to derive the luminosity function for this young cluster will be described and the results compared to that observed for visible open clusters. A short discussion will be devoted to the energy balance of the cloud in light of new far-infrared observations.

II. Molecular Cloud Structure

In a survey of $^{12}C^{16}O$ emission from dark clouds, Penzias et al. (1972) noted that the dark cloud which lies near the star ρ Ophiuchi had an elevated temperature relative to the other dark clouds in the survey. Subsequent observations by Encrenaz (1974) located strong molecular emission from the western region of the dark cloud complex. Since these pioneering observations, detailed mapping of molecular emission lines have continued to reveal new information concerning the gas temperature structure, column density distribution and high-density gas in the core of the ρ Oph cloud.

A. Gas-Temperature Structure

Detailed mapping of the ^{12}CO (1-0) emission lines in the central regions of the ρ Oph cloud were first performed by Encrenaz, Falgarone, and Lucas (1975). More extensive maps followed with higher velocity resolution of the narrow lines (FWHM ~2-4km s^{-1}), producing a clear overview of the gas temperature structure of the molecular gas (Loren *et al.* 1980). As shown in Fig. 1, the strongest CO emission (T_A^* ~40-50K) is found at the western edges of the cloud in the vicinity of the B2V star HD147889 and the B9-A0V star SR-3. While the gas temperatures appear to drop continuously to the east of these stars, CO emission lines also become heavily self-absorbed by cold foreground gas toward the dense core. As a result, self-absorption conceals the detailed distribution of the gas temperature in the central regions of the cloud.

B. Column Density Distribution

The column density of molecular gas is ideally computed from observations of optically-thin emissions lines of the isotopes of CO by assuming a local thermodynamic equilibrium model and a common excitation temperature for all isotopes of CO. An extensive preliminary map of ^{13}CO emission lines in the ρ Oph cloud by Loren (1984) is shown in Fig. 2 and traces out the core-streamer morphology of the dark cloud complex. Unfortunately, ^{13}CO emission lines are optically-thick toward the denser regions at the western edge of the ρ Oph complex

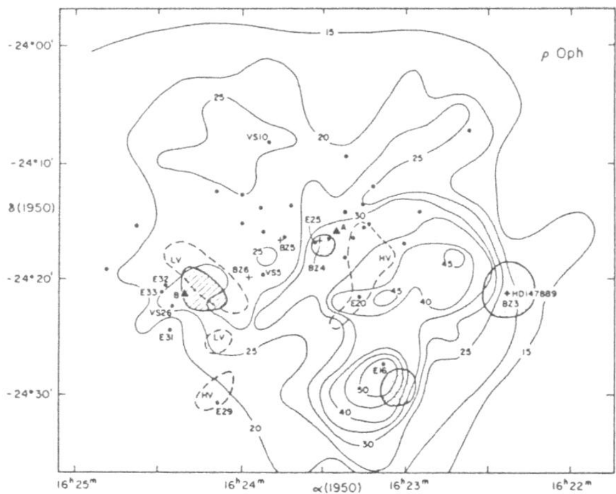

Fig. 1 - A map of T_A^*(^{12}CO 1-0) in the core of the ρ Oph cloud by Loren et al. (1980). The peak antenna temperatures are closely associated with dust emission (shown by dotted areas) excited by the B stars HD 147889 and SR-3 (E16).

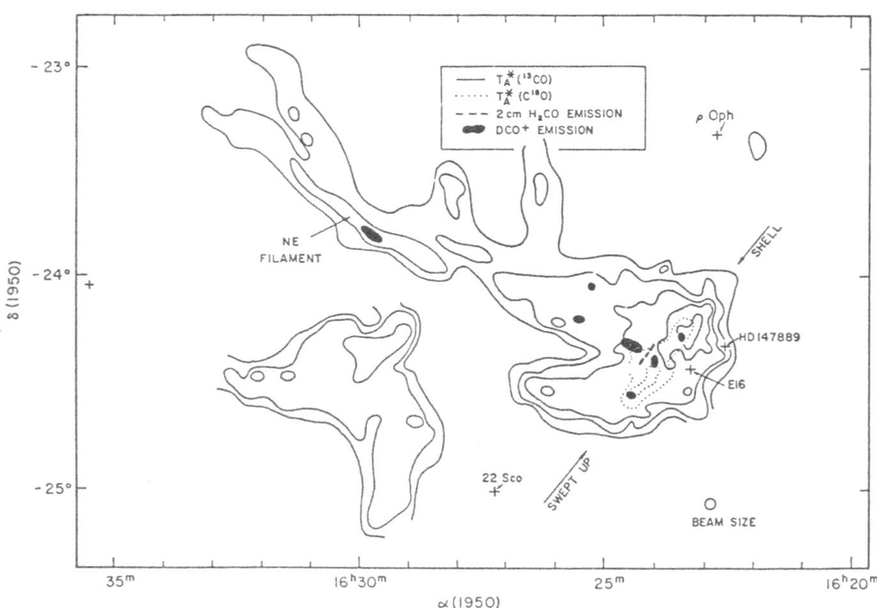

Fig. 2 - The column density structure of the ρ Oph dark cloud complex as presented by Loren (1984). The solid contours represent ^{13}CO emission with contour intervals of T_A(^{13}CO) = 3,6,10,15, and 20K. The dotted contours trace out the high column density cloud core as defined by C^{18}O emission. The molecular compression front proposed by Loren and Wootten (1984) is delineated by the 2 cm H$_2$CO emission while the cold pre-shock gas is represented by the DCO$^+$ peak to the northeast.

as evidenced by their self-absorbed profiles (Lada and Wilking 1980). The best tracer of the molecular column density in the ρ Oph cloud core is the $^{12}C^{18}O$ emission.

Strong $^{12}C^{18}O$ (1-0) emission lines are observed in the ρ Oph cloud and have been used to delineate the 1 pc x 2 pc ridge of gas which forms the centrally condensed core (see Fig. 3 and Wilking and Lada 1983). The gas column densities in this core are extremely high and range from $N^{18}_{LTE} \sim 1.5\text{-}3.0 \times 10^{16}$ cm^{-2} and imply visual extinctions of $A_V \sim 50 - 100$ mag. As discussed in future sections, the distribution of this high column density gas laid important groundwork for infrared observations which studied the population of young stars embedded in the dense core and for our understanding of the energy balance of the cloud.

C. High Density Gas

Emission from molecules which require densities in excess of $n(H_2) = 10^4$ cm^{-3} to excite have been mapped throughout the high column density core. Surprisingly, the emission from these density sensitive molecules is not necessarily correlated with the highest column densities of gas. Observations of SO and H_2CO show that highest density gas ($n(H_2)>10^5$ cm^{-3}) is collected into two major concentrations referred to as "ρ Oph A" and "ρ Oph B" (Gottlieb et al. 1978, Loren et al. 1980, Loren, Sandqvist and Wootten 1983). As marked by the 'X's in Fig. 3, ρ Oph A is located at northern extent of the high column density ridge of $C^{18}O$ while ρ Oph B is situated to 15 arcmin to the southeast and adjoins ρ Oph A by a plateau of dense gas ($n(H_2)>10^4$ cm^{-3}). ρ Oph B lies just east of the high column density ridge of gas and evidently occupies a much smaller fraction of the total gas column than the dense gas toward ρ Oph A.

The dense core ρ Oph B can be further subdivided in fragments which are characterized by emission from NH_3 (Zeng et al. 1984), DCO^+ (Loren and Wootten 1984) and rare emission from H_2CO at 2 cm (Loren et al. 1980). The distribution in space and velocity of these dense clumps are described by Loren and Wootten (1984) in the context of a model for a compression front in the ρ Oph B area. This compression front would have propagated from the southwest and perhaps initiated star formation in the swept up high column density ridge.

III. Infrared Observations of Star Formation

Early emission-line surveys of the ρ Oph region had revealed a number of Hα emission objects and T Tauri stars (e.g., Struve and Rudkjøbing 1949, and Dolidze and Arakeylyan 1959). However, the most effective studies of the large population of young stars embedded in the ρ Oph cloud have been through infrared observations of these heavily obscured objects. By combining near-infrared surveys with broadband photometry from 1-100 μm, we are able to obtain insight into the number and nature of stars which have recently formed in the core of the ρ Oph cloud.

A. Near-Infrared Surveys

One outstanding feature of star formation in the ρ Oph cloud is the high density of 2 μm sources observed toward the core. Once corrected for contamination by field stars, the density of young stars in the ρ Oph core is by far higher than that in any other nearby cloud complex. Four major 2 μm surveys have sampled the entire extent of the dense core with varying degrees of coverage and sensitivity. From these surveys, nearly 50 young stars have been identified within a radius of 2 pc from the center of the core (marked by open triangles in Fig. 3).

The first near-infrared survey in the ρ Oph cloud was performed by Grasdalen, Strom, and Strom (1973) to a limiting magnitude of K = 9.0. This was soon followed by an extensive survey by Vrba et al. (1975) with a sensitivity of K = 10 mag. While the nearly 70 two-micron sources discovered by these surveys indicated the presence of a young infrared cluster, it was necessary to correct the population for background contamination before investigating the true size and nature of the embedded cluster. A subsequent 2 μm survey by Elias (1978) was aimed at sorting out field stars from the sample. The primary criterion used by Elias was the 2.3 μm absorption feature present in the background giant stars but absent or obscured by dust toward the embedded young stars. Most recently, a 10' x 10' area enclosing the central portion of the $C^{18}O$ ridge was surveyed to a sensitivity of K = 12 mag by Wilking and Lada (1983). Twenty sources were revealed by this deep survey (16

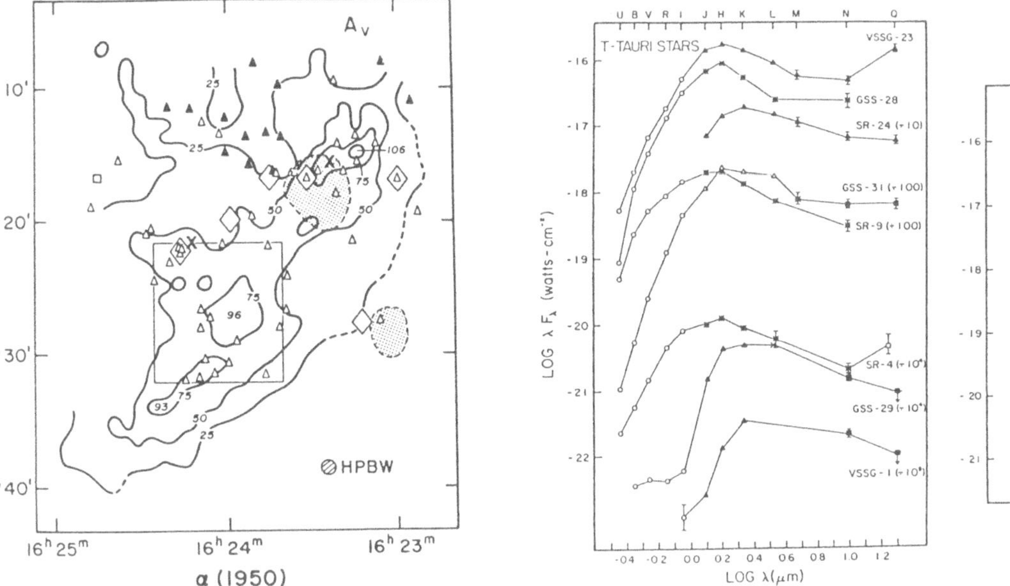

Fig. 3 - The centrally condensed core of the ρ Oph cloud as defined by $C^{18}O$ emission (Wilking and Lada 1983). The 1 pc X 2 pc core is shown by contours of A_v which are inferred from the $C^{18}O$ column density. Infrared sources which are members of the young cluster are denoted by open triangles, unclassified sources by solid triangles. Regions of far-infrared emission measured by Fazio et al. (1976) are shown by dotted areas and the areas of dense gas which mark the location of ρ Oph A and ρ Oph B are indicated by X's (Loren et al. 1980).

Fig. 4 - A sampling of spectral energy distributions for young stars embedded in the ρ Oph cloud taken from Lada and Wilking (1984). Fig. 4a show the energy distributions for eight stars proposed to be T Tauri stars. Their energy distributions are characteristic of most infrared sources in ρ Oph; they are much broader than a single temperature blackbody and suggests the presence of excess emission from hot circumstellar dust. In contrast, Fig. 4b shows 3 stars whose energy distributions can be fit with a single temperature reddened blackbody for $\lambda < 10\mu m$.

were newly discovered) and all were shown to be embedded in the cloud on the basis of their infrared excesses at 3.4 μm or their superposition on regions of high extinction.

B. The Nature of the Embedded Stars

Broadband ground-based (1-20 μm) and far-infrared (40-160 μm) observations have been critical for the study of the nature of the ρ Oph embedded infrared cluster. Without the benefit of extensive photometry, the luminous near-infrared colors of the cluster members were first interpreted as arising from the reddened photospheres of early-type stars. However, the presence of a large, luminous cluster did not agree with the absence of strong far-infrared emitting dust. Fazio et al. (1976) found only three sources of dust emission in the ρ Oph core which were excited by early-type stars. The resolution to this problem was provided by Elias (1978) and Lada and Wilking (1984) who showed that the majority of the ρ Oph cluster members display broad energy distributions in the 1-20 μm spectral region characteristic of young stars surrounded by circumstellar dust (see Fig. 4). Many of these stars are probably still contracting toward the main sequence. The steeply rising energy distributions of these stars toward longer wavelengths are not due to the reddening of a luminous stellar photosphere but arise from a reddened photosphere of a lower luminosity star plus emission from hot circumstellar dust.

Indeed, subsequent mid-infrared and far-infrared observations of the ρ Oph cloud have shown that is has formed a predominantly low-luminosity cluster. Lada and Wilking (1984) have presented a detailed study of the 1-20 μm spectral energy distributions for 32 of the ρ Oph cluster members. They find that most of the sources emit the bulk of their energy in the 1-20 μm spectral region. Many of these stars have energy distributions which resemble those of reddened T Tauri stars (ref. Fig. 4). In most cases luminosities could be computed reliably (within a factor of 2) for these sources by simply integrating their observed energy distributions. By computing the luminosities in this manner, it is not necessary to deredden the energy distribution since radiation absorbed at shorter wavelengths is presumably recovered at the mid-infrared wavelengths.

Nearly half (44%) of the sources studied by Lada and Wilking (1984) were computed to have bolometric luminosities less than the sun (ref. Fig. 5). Apart from three B stars, the ρ Oph cluster is comprised entirely of low mass stars. The B stars provide nearly all of the cluster luminosity of 6500 $L_\odot$. Recent far-infrared observations (Sargent *et al.* 1983, Cudlip *et al.* 1984, Wilking *et al.* 1984) which have been sensitive to embedded stars of 10-20 $L_\odot$ support this picture of a low-luminosity cluster.

IV. The Formation of a Bound Cluster

With a clearer picture of the spatial distribution of young stars, their luminosities, and the molecular environment in which they are embedded, more recent investigations have been able to concentrate on more global aspects of star formation in the ρ Oph cloud. Because of the high density of young stars, it had been proposed that the ρ Oph cloud is the site of a young open cluster in the process of formation. To make this comparison more quantitative one can compute the current efficiency of conversion of gas into stars or star formation efficiency (SFE). The SFE is a time-dependent quantity defined as $M_{stars}/(M_{stars} + M_{gas})$. Final values for the SFE must exceed 25% to 50% to produce a gravitationally bound cluster depending on whether the gas is removed from the cluster slowly (4-5 x 10^6 years) or rapidly ($<10^6$ years) (Lada, Margulis, and Dearborn 1984).

The current best estimate for the SFE in the core region of the ρ Oph cloud is 25% (Lada and Wilking 1984). This estimate assumes two populations of stars within a star-forming layer with a depth corresponding to 40 mag of extinction: one population is comprised of 0.5 $M_\odot$ stars which have an apparent brightness at 2.2 μm of K = +10 at the cloud surface and the second consists of solar mass stars with an apparent brightness of K = +8.0 mag at the cloud surface (the brightness of a typical one solar mass T Tauri star). While the SFE is a difficult parameter to estimate, the value of 25% has been arrived at conservatively by using two populations of low mass stars and by computing the mass of molecular gas directly through the gas-to-dust ratio.

The high value for the SFE and the absence of massive stars suggest the ρ Oph cloud core is in the process of forming what will emerge as a gravitationally bound open cluster. The lack of disruptive stars guarantees the continued production of stars and an increase in an already high conversion efficiency of gas into stars. In addition, the quiescent cloud conditions should permit the slow release of gas from the system of gas and stars and relax the requirements on the final SFE needed to produce a bound cluster. The stellar mass density of ~145 $M_{\odot}$ pc^{-3} which is estimated for the ρ Oph core underlines the stability of the young embedded cluster against disruption from galactic tides and passing interstellar clouds.

V. The Luminosity Function of Embedded Stars

The key to our understanding of how open clusters form and evolve lies in the study of young clusters still embedded in their molecular clouds. In particular, the luminosity function of an embedded cluster, when compared to those of gas-free open clusters, can reveal how the mass function of the cluster evolves during the formation process and how protostellar fragments are produced to form stars. Such details about star formation are not easily accessible from studies of visible clusters or of individual protostars alone.

The luminosity function of a zero-age-main sequence open cluster is a well-defined distribution which varies insignificantly from cluster to cluster and closely resembles the "initial" luminosity function derived from field stars (see Scalo 1984 for review). The basic similarity of these initial luminosity functions are surprising in light of the extended episodes of star formation (10^7-10^8 years) suggested for several open clusters (Herbig 1962, Landolt 1979, Stauffer 1980, Herbst and Miller 1982). Even more startling is the NGC 2264 cluster which appears to have formed stars sequentially in mass (starting with the lowest mass stars) over 10^7 years and yet displays a luminosity function which resembles the initial luminosity function derived from field stars (Adams, Strom, and Strom 1983).

From their study of the spectral energy distributions of the ρ Oph cluster members, Lada and Wilking (1984) have assembled the first luminosity function for an embedded cluster. This luminosity function is shown in Fig. 5a for 37 cluster members with well-determined luminosities. For comparison, the luminosity function corresponding to a log-normal initial mass function (Miller and Scalo 1979) is also plotted and normalized to the observed low-luminosity population. The deviation of the observed luminosity function from that of the initial mass function (IMF) at low luminosities reflects the sensitivity limits of infrared surveys and there are no other statisically significant deviations.

However, before a meaningful comparison of the ρ Oph cluster luminosity function can be made to that from the IMF, all known selection effects must be considered. The most important of these is (1) the inability of infrared surveys to sample the embedded population throughout the entire depth of the cloud and (2) the variation of the cloud depth which is completely sampled with luminosity (Lada and Wilking 1984). As shown in Fig. 5b, correcting for the incomplete sampling of infrared surveys suggests there may be a deficiency of intermediate-to-high luminosity ($L>5L_{\odot}$) stars in the ρ Oph cluster relative to the IMF.

Any proposed deviations of the ρ Oph cluster luminosity function from that of the IMF must be confirmed through the observation of additional cluster members and the far-infrared luminosities of heavily obscured or cool stars. The subsequent evolution of the young embedded cluster to the main sequence must also be properly accounted for in the observed luminosity function. However, if confirmed, the deficiency of intermediate-to-high luminosity stars in the ρ Oph cluster would have a great impact upon our understanding of the star formation process. If the ρ Oph cloud core is ultimately to form a cluster with an "initial" luminosity function which is common to that observed in visible open clusters, then future episodes of star formation would have to produce more massive stars. This type of evolution for an embedded cluster's mass function has been suggested for the NGC 2264 cluster (Adams, Strom, and Strom 1983). Further study of the luminosity function of embedded clusters may also give us insight into how protostellar fragments are produced. For example, while the present ρ Oph luminosity function suggests a general deficiency of stars with $L>5L_{\odot}$, two luminous B stars, Source 1 (1500 $L_{\odot}$)

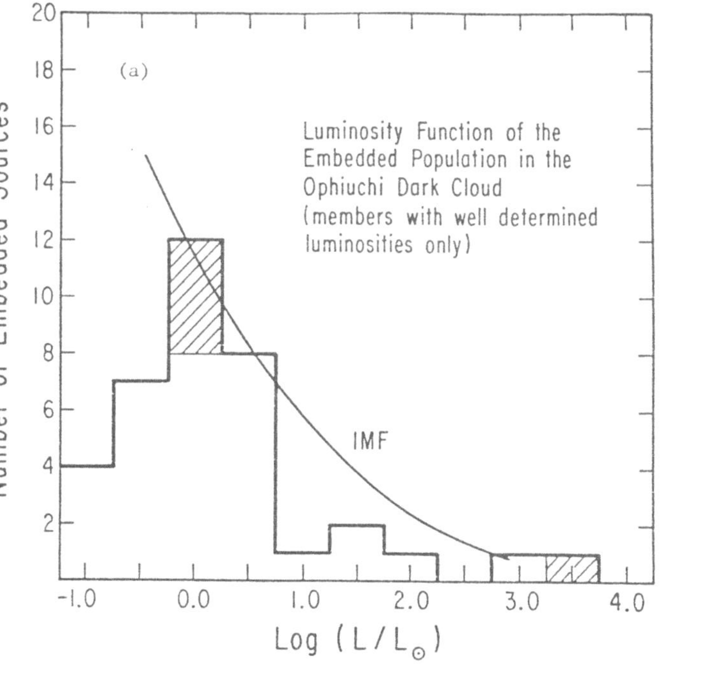

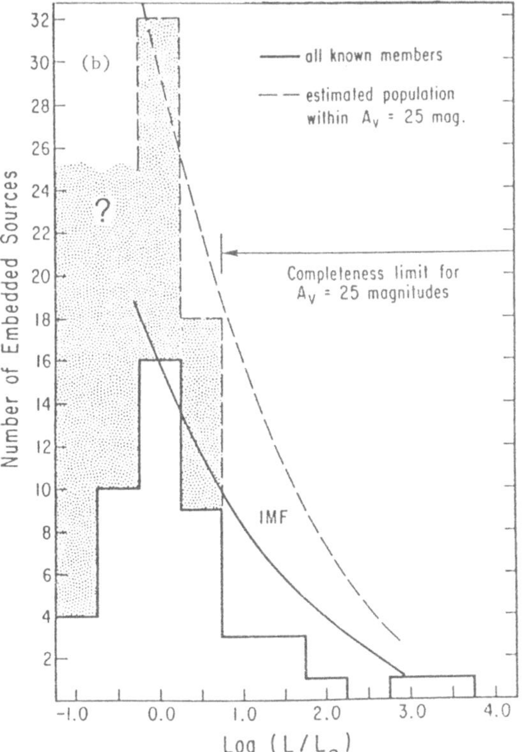

Fig. 5 - The luminosity function of the Oph cluster presented by Lada and Wilking (1984). Fig. 5a shows the observed luminosity function for 37 stars with well-determined luminosities. The luminosity function corresponding to a log-normal IMF (Miller and Scalo 1979) is also shown, normalized to the low-luminosity stars. Fig. 5b presents the luminosity function for all 48 members of the ρ Oph cluster and the shaded area indicates the correction to this population which needs to be made to account for incomplete sampling by infrared surveys. Only when selection effects are accounted for is there a significant deviation of the luminosity function from that derived from the IMF at $L \geqslant 5 L_{\odot}$.

and HD 147889 (5000 $L_\odot$), have also formed in the core region. Their presence is reminiscent of the bimodal mass distribution predicted by models for the growth of protostellar fragments by accretion (Zinnecker 1982) and by collisional coalesence (Pumphrey and Scalo 1984). In these models, the growth mechanisms lead to a runaway growth of the most massive fragment and a gap in the resulting mass function.

VI. The Cloud Energetics

Ten years ago, the energy balance of the ρ Oph cloud was understood in terms of a luminous cluster of early-type stars embedded in a warm molecular cloud. However, to complete this simple picture for the cloud energetics required the presence of luminous emission from warm dust. The dust would be radiatively heated by the young stars which, in turn, would heat the molecular gas via collisions in the dense cloud. The failure to detect large quantities of warm dust associated with this young cluster by Fazio et al. (1976) and by subsequent far-infrared studies cast doubt on the standard model for cloud heating in ρ Oph. The situation was made worse by the discovery of the extremely large columns of gas in the core by Wilking and Lada (1983). Surely, a column of dust with a visual extinction of 50-100 mag would be easily observed if the dust temperature was greater than or equal to the observed gas excitation temperatures derived from ^{12}CO of 35-50K. The identification of the ρ Oph cluster as a low-luminosity cluster by Lada and Wilking (1984) dominated by only two early B stars ($L_{total} \sim 6500\ L_\odot$) completed the demise of the simple model of the energy flow in the ρ Oph cloud.

There are at least three separate approaches which might be taken to resolve the presence of warm gas in ρ Oph with the absence of warm dust. The first is to explore heating mechanisms which may be operating in the vicinity of ρ Oph which might deliver energy directly to the gas, bypassing the dust. These mechanisms may include shock heating, the subsequent release of stored magnetic energy from a compressed cloud or cosmic-ray heating. Shocks have been proposed to have traversed the ρ Oph cloud based on the cloud morphology, magnetic field geometry and the high density of 2 μm sources (Vrba 1977). A low velocity compression front within the core of the cloud has been suggested to explain the velocity structure and cloud chemistry in the dense ρ Oph B region (see Fig. 2 and Loren and Wootten 1984). Shocks which would be associated with the North Polar Spur supernova remnant have been forwarded to explain a gamma-ray source observed toward the cloud (Morfill et al. 1981). The same gamma-ray source, if identified within the ρ Oph cloud, could provide an enhanced cosmic-ray flux which could heat the molecular gas in ρ Oph to the observed temperatures. However, the observed gamma-ray flux may be consistent with the interaction of ambient cosmic rays with the molecular cloud material (Issa, Strong, and Wolfendale 1981).

A second avenue of study which may shed light on the energetics problem in the ρ Oph cloud is to investigate the importance of extended far-infrared emission in the cloud. The 100 μm IRAS images of the ρ Oph (shown by Beichman at this conference) display extended emission from warm dust throughout the ρ Oph cloud; this extended component would have been subtracted out by previous airborne and balloon-borne observations. It will be important to learn the physical temperature of this warm dust and see if the gas density in these outlying areas of the cloud are high enough ($n(H_2)>5 \times 10^4$ cm^{-3}) to provide sufficient dust-gas coupling to heat the molecular gas via collisions.

A final resolution to the energy balance in the ρ Oph cloud may simply involve investigating the temperature structure of the molecular gas. Since ^{12}CO emission lines can saturate within a relatively low column density, the warm gas temperatures observed in ρ Oph could arise merely from a surface layer of the high column density core. The heating of this surface layer could be supplied by the existing B stars. Indeed, the high degree of correlation between far-infrared emission and ^{12}CO emission line strength suggests they have a similar heating source (see Fig. 2b, Cudlip et al. 1984). Perhaps multi-transition studies of specific molecules coupled with the continued observation of density-sensitive molecules may unravel the temperature structure of the gas and shed light on the energy balance of the ρ Oph cloud.

Acknowledgments

I am grateful to the Organizing Committee of the Nearby Molecular Cloud symposium for their hospitality and to the American Astronomical Society for an International Travel Award. I also thank Bob Loren for data in advance of publication.

References

Adams, M.T., Strom, K.M., and Strom, S.E. 1983, Ap. J. Suppl., 53, 893.
Cohen, M. and Kuhi, L.V. 1979, Ap. J. Suppl, 41, 743.
Cudlip, W., Emerson, J.P., Furniss, I., Glencross, W.M., Jennings, R.E., King, K.J., Lightfoot, J.F., and Towlson, W.A. 1984, M.N.R.A.S., in press.
Dolidze, M.V., and Arakeylyan, M.A. 1959, Soviet Astr. - AJ, 3, 434.
Elias, J.H. 1978, Ap. J., 224, 453.
Encrenaz, P.J. 1974, Ap. J. (Letters), 189, L135.
Encrenaz, P.J., Falgarone, E., and Lucas, R. 1975, Astr. Ap., 44, 73.
Fazio, G.G., Wright, E.L., Zielik, M., III, and F.J. Low, 1976, Ap. J. (Letters), 206, L165.
Gottlieb, C.A., Gottlieb, E.W., Litvak, M.M., Ball, J.A., and Penfield, H. 1978, Ap. J., 219, 77.
Grasdalen, G.L., Strom, K.M., and Strom, S.E. 1973, Ap. J. (Letters), 184, 736.
Herbig, G.H. 1962, Ap. J., 135, 736.
Herbst, W., and Miller, D.P. 1982, A.J., 87, 1478.
Issa, M.R., Strong, A.W., and Wolfendale, H.W. 1981, J. Phys. G: Nucl. Phys, 7, 565.
Lada, C.J. and Wilking, B.A. 1980, Ap. J., 238, 620.
Lada, C.J., and Wilking, B.A. 1984, Ap. J., in press.
Lada, C.J., Margulis, M., and Dearborn, D. 1984, Ap. J., 285, 141.
Landolt, A.U. 1979, Ap. J., 231, 468.
Loren, R.B. 1984, Personal communication.
Loren, R.B., Wootten, A.W., Sandqvist, Aa., and Bernes, C. 1980, Ap. J. (Letters), 240, L165.
Loren, R.B., Sandqvist, Aa., and Wootten, A.W. 1983, Ap. J., 270, 620.
Loren, R.B. and Wootten, A.W. 1984, preprint.
Miller, G.E., and Scalo, J.M. 1979, Ap. J. Suppl., 41, 513.
Morfill, G.E., Volk, H.J., Drury, L., Bignami, G.F., and Carareo, P.A. 1981, Ap. J., 246, 810.
Penzias, A.A., Soloman, P.M., Jefferts, K.B., and Wilson, R.W. 1972, Ap. J. (Letters), 174, L43.
Pumphrey, W.A., and Scalo, J.M. 1984, Ap. J., in press.
Rydgren, A.E., Strom, S.E., and Strom, K.M. 1976, Ap. J. Suppl., 30, 307.
Sargent, A.I., Van Duinen, R.J., Nordh, H.L., Fridlund, C.V.M., Aalders, J.W.G., and Beintema, D. 1983, A.J., 80, 88.
Scalo, J.M. 1984, Fundamentals of Cosmic Physics, in press.
Stauffer, J.R. 1980, A.J., 85, 1341.
Struve, O., and Rudkjøbing, M. 1949, Ap. J., 109, 92.
Vrba, F.J. 1977, A.J., 92, 198.
Vrba, F.J., Strom, K.M., Strom, S.E., and Grasdalen, G.L. 1975, Ap. J., 197, 77.
Wilking, B.A., and Lada, C.J. 1983, Ap. J., 274, 698.
Wilking, B.A., Harvey, P.M., Joy, M., Hyland, A.R. and Jones, T.J. 1984, Ap. J., in press.
Zeng. Q., Batrla, W., and Wilson, T.L. 1984, Astr. Ap., in press.
Zinnecker, H. 1982, in Symposium on the Orion Nebula to Honor Henry Draper, ed. Glassgold, Huggins, and Schucking, (New York: New York Academy of Sciences), 226.

WARM DUST IN THE RCRA MOLECULAR CLOUD

A. Leene
Kapteyn Laboratory
PO Box 800
9700 AV Groningen
The Netherlands

C.A. Beichman
Department of Physics
CalTech
Pasadena, CA 91125
USA

Abstract:

We report on IRAS observations of a portion of the RCrA molecular cloud. The observations show IR emission coinciding with an area of low optical extinction. The 60-100 μm color temperature of this dust is 21 K. The 12-25 μm color temperature is however 400 K. The emission of this warm dust extends over an area of 0.3 square degrees. This high temperature can not be explained by normal equilibrium processes. The only available heating source, the Inter Stellar Radiation Field (ISRF), can heat the dust to about 20 K. The observations imply the existence of very small grains, which are heated by single photon absorptions.

IRAS maps at 12, 25, 60 and 100 μm of a 1 by 1 degree field in the R Corona Australis molecular cloud are presented. The measurements were taken from the All Sky Survey (Neugebauer et al(1984)), which scanned the area 6 times. The accuracy of the absolute calibration is 5% at 12 and 25 μm and 10% at 60 and 100 μm. The beamsizes are .75 x 4.5, .75 x 4.6, 1.5 x 4.7 and 3.0 x 5.0 arcmin for 10, 25, 60 and 100 μm respectively. The beam position angle is -7 degrees. The data are shown in figure 1.

At all wavelengths an IR ridge bends around a star at $18^h53^m17.1^s$ and $-37^o24'33''$ (HR 7129). The maximum intensity is formed at a point closest to this star. The 60 and 100 μm pictures are quite similar. The IR emission extends into the main cloud to the east, which lies partly outside the present map. However the 12 and 25 μm emission is quite different in appearance. Here the emission seems to be seperated from the main cloud to the east. From the optical plates it is clear, that the 60 and 100 μm emission correspond very well with the optical extinction. The 12/100 μm flux ratio on the ridge is about 1, whereas near the main cloud to the east it is 0.15. The extinction in the main cloud to the east is also higher than on the ridge. The significance of this is not yet clear.

The relevant data have been presented in Table I. The total flux density for the 12 and 25 μm maps has been calculated by integrating the brightness above the appropriate 5 sigma levels. For the 60 and 100 μm maps an apropriate cutoff has been chosen at a local minimum on the ridge. We also indicate the brightnesses for the 100 μm peak, as well as the νF_ν brightnesses, to give an idea of the spectrum. The corresponding color temperatures for a constant dust emissivity have been listed. In brackets the dust temperature for a λ^{-2} emissivity is listed.

Table I : Relevant data

wave-length	total flux density		Bright-ness	νF_ν	T_{color}
μm	10^{-12}W/m^2	Jy	10^{-7}W/m^{2}ster	10^{-7}W/m^{2}ster	Kelvin
12	14.1	112	4.0	7.9	
					360
25	4.19	84.2	1.1	2.7	
					100
60	6.01	237	1.8	3.5	
					30 (21)
100	12.7	1250	2.9	8.6	

The spectrum in νF^ν shows a surprising minimum at 25 and 60 μm, while the 12 and 100 μm fluxes are almost equal. This can also be seen in the color temperatures, which increase from the long to the short wavelengths. The 12 and 25 μm fluxes are several orders of magnitude larger than is expected from the 60 and 100 μm extrapolation.

From the optical plates the most obvious heating source is the star HR 7129. Extended reflection nebulosity is seen around the star indicating that it is in interaction with the surrounding dust. There is, however, no evidence at all for this nebulosity in the IR observations. The distance of 244 pc (Guttierrez-Moreno & Moreno(1968)) to the star is also larger than the distance to the RCrA cloud, which is 130 pc (Marraco & Rydgren(1981)). It is therefore likely that two different dust clouds are involved. One cloud is responsible for the nebulosity around the star, at 244 pc, and the other is part of the RCrA cloud. The nebulosity around the star is only very thin and has not enough column density to be visible in the IR. No other local sources, which can heat the dust, seem to be available. These would have been readily visible on the optical plates, where the extinction is only 1 magnitude. On the IR maps these embedded heating sources would also be easily visible. None are, however, observed.

The only other available source is the Interstellar Radiation Field. This radiation field can heat the dust to temperature of about 20K for graphites or about 15K for silicates(Mezger et al(1982)). This is much smaller than the observed 400K. If we calculate the total mass involved in the 60 and 100 μm emission using formula 12 of Mezger et al(1982) a value of about 1 $M_\odot$ is expected, which is not an unlikely amount. Also the 100 μm flux is expeceds from heating by the ISRF (assuming A_V=1).

The main problem is thus the 12-25 μm color temperature. This can not be explained by normal equilibrium processes. A possible solution lies in the thermal fluctuation model e.g. Sellgren(1984) and references therein. In this model very small grains are transiently heated to very large temperatures by absorption of a single photon. The current observations would then imply the existence of very small grains with sizes of a few Å. This is much smaller than the minimum size assumed in the MRN(Mathis et al(1977)) grain size distribution, which is 50 Å. Sellgren observed the coexistence of a large color temperature and the unidentified emission feature at 3.3 μm. Leger & Puget(1984) proposed large polycyclic aromatic molecules for the explanation of the unidentified IR features. This could provide a link between these features and the small grains. These small grains do not radiate like a blackbody, but radiate predominantly in these features. One of these features lies at 11.3 μm, well within the 12 μm band. This feature can also contribute to the observed 12 μm excess.

A detailed analysis of these suggestions is still in progress. It is, however, clear that the need for the small grains exists and that they are possibly responsible for the unidentified features. From the IRAS data the existence of these grains becomes more evident (see the other articles on IRAS in this volume). The suggestion that the features are only present when there is a UV source present, as has been suggested by Sellgren, no longer seems valid. In the diffuse medium these features are also present. Ground based observations are, however, not able to detect these lines in the diffuse medium, because of sky chopping and a lack of sensitivity.

References:

Leger & Puget(1984), Astron, Astrophys. 137:L5
Marraco & Rydgren(1981) Astron. J. 86:62
Mathis et al(1977), Astroph. J. 217:425
Mezger et al(1982), Astron. Astrophys. 105:372
Guttierez-Moreno & Moreno(1968), Astroph. J. Supp. 15:459
Neugebauer et al(1984), Astroph. J. 278:L1
Sellgren(1984), Astroph.J. 277:623

Figure 1: These plots represent the 12, 25, 60 and 100 µm emission. The contours lie at .6, .9, 1.2, 1.5, 2, 2.5, 3, 3.5, 4, 4.5 and 5 10^{-7}W/m^2ster. The crosses represent stars from the SAO catalog.

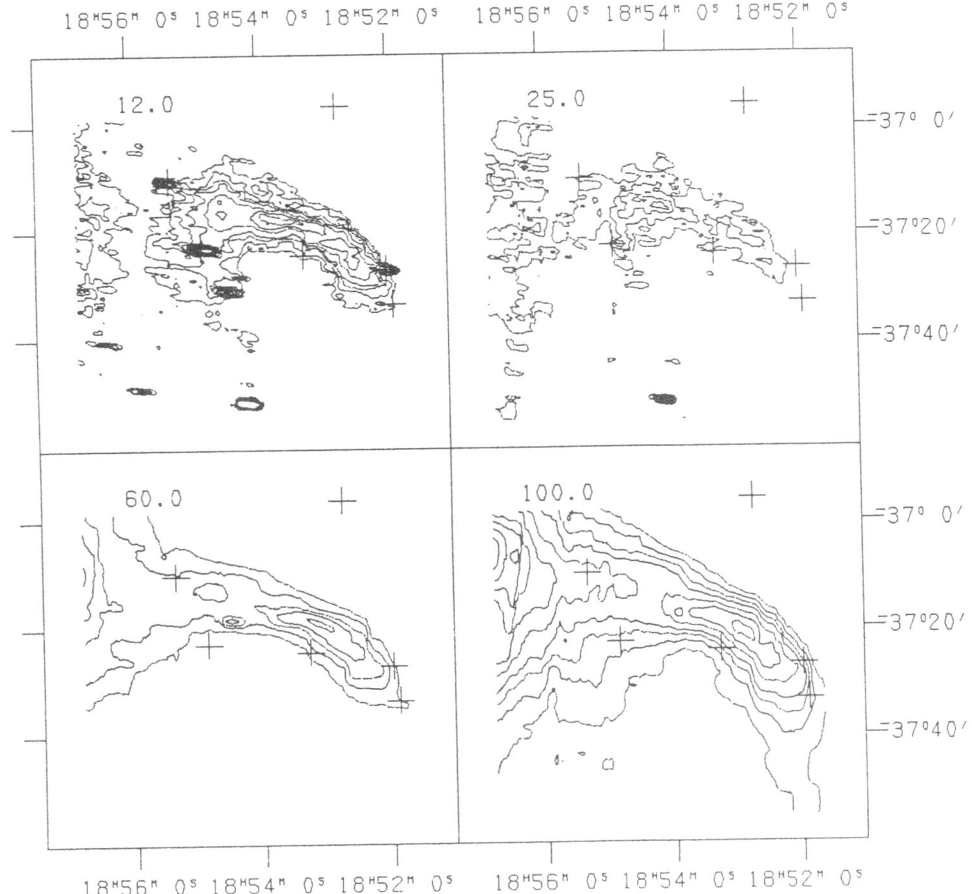

PART IV

MISCELLANEOUS RELATED TOPICS

ON THE POSSIBLE CONTRIBUTION OF THE FRUSTRATED TOTAL REFLECTION IN THE COMPOSITE DIELECTRIC GRAINS TO THE EXTINCTION AND POLARIZATION OF LIGHT

G.B. SHOLOMITSKII
Space Research Institute
Academy of Sciences of the USSR
MOSCOW

Given the dielectric nature of interstellar grains (Martin 1973) and the presence of materials with different refractive indices within the dust-molecular clouds (silicates and ice, e.g. Willner et al., 1982) the core-mantle grains considered in detail by Greenberg (1982) may form by the coagulation the composite grains with the lower refractive index optical gaps in the bulk material. Then, the frustrated total reflection (FTR) characteristic for such composite structures will manifest the specific λ-dependence of the extinction and prolongation that can be used to distinguish coagulated population of grains.

Let us assume for the first qualitative consideration that the optical gap with the thickness d and refractive index $n_2 < n_1$ in the bulk material is optically symmetric ($n_3 = n_1$) and has both longitudinal and transverse dimensions large enough $l \gtrsim (1 - 2)\ \lambda$, $2a \gtrsim 4\ \lambda/\pi$ respectively for having reflected wave formed for the most values of the incidence angle and for FTR prevailing over diffraction. Then, for the incidence angles larger than the critical value $\theta_{cr} = \sin^{-1}(n_2/n_1)$ the transmission of the gap for the two orthogonal polarizations is given by (Hall 1902) :

$$T_s = \frac{4\, n^2 \cos^2\theta\, (n^2 \sin^2\theta - 1)}{(n^2-1)^2 \sinh^2\mu + 4\, n^2 \cos^2\theta\, (n^2\sin^2\theta - 1)}$$

$$T_p = \frac{4\, n^2\cos^2\theta\, (n^2\sin^2\theta - 1)}{(n^2-1)^2\, (n^2\sin^2\theta - \cos^2\theta)^2 \sinh^2\mu + 4n^2\cos^2\theta\, (n^2\sin^2\theta - 1)} \qquad (1)$$

where s and p mark polarizations perpendicular and parallel to the incidence plane, $n = n_1/n_2$,

$$\mu = \frac{2\pi d\, n_2}{\lambda} \sqrt{n^2\sin^2\theta - 1}$$

and reflection coefficients are $R_{s,p} = 1 - T_{s,p}$. It is immediatly seen from (1) that transmission for both polarizations varies in the range 0 - 1 depending on the wavelength due to hyperbolic sine in the denominators and that transmitted light becomes polarized because of the unequality $T_s \neq T_p$.

To look at pure FTR effects we ignore at present the Frenel reflection that is neutral and occurs only for the small incidence angles $\theta < \theta_{cr}$. For the composite grains having the optical gaps nearly perpendicular to the axes and perfectly aligned in the magnetic field by the mechanism of the Davis-Greenstein type their relative amount showing FTR is

$$f = 1 - \frac{2}{\pi} \cos^{-1} (\sqrt{1 - 1/n^2}/\sin \gamma_B), \quad (2)$$

where γ_B is the inclination of the magnetic field to the line of sight. It increases with n and decreases with γ_B being equal f = 1 for magnetic field inclination $\gamma_B < 90° - \theta_{cr}$ (Fig. 1).

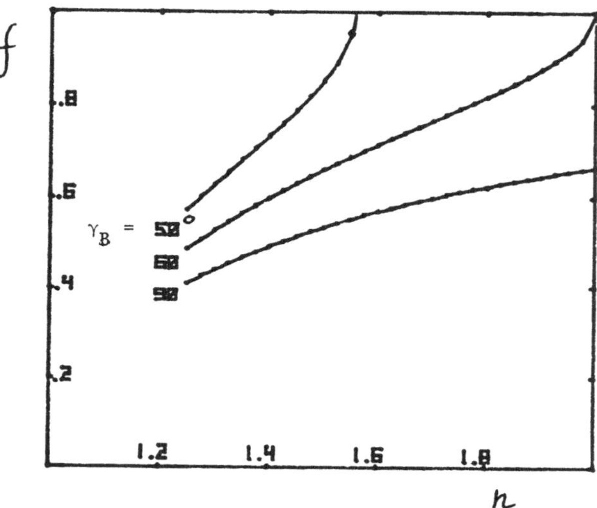

Fig. 1 The proportion of composite grains with optical gaps perpendicular to the axis aligned by the familiar magnetic mechanism to show FTR.

Using for the extinction and polarization the relationships

$$\Delta m_\lambda = 1.086 \text{ NGL} \times \frac{1}{2} < (R_s + R_p) \cos\theta > ,$$

$$P_\lambda = \tanh\left(\frac{\Delta\tau}{2}\right) \simeq \text{NGL} \times \frac{1}{2} < (T_s^B - T_p^B) \cos\theta > \quad (3)$$

and integrating with (1) over incidence angles $\theta > \theta_{cr}$ we obtain for the expected contribution of FTR to the extinction and polarization and for the ratio $P_\lambda / \Delta m_\lambda$ the wavelength dependences shown in Fig. 2 and 3. On (3), G is the typical geometrical area of the gaps, $\frac{1}{2} < (R_s + R_p) \cos\theta >$ is equivalent to the usual effectivity (Q_e), $\Delta\tau$ is the opacity difference for two polarizations across and along magnetic field (designated by the upper subscript B) and the gaps are assumed to be uniformly distributed within the angle $\Psi = \sin^{-1}(\sin\alpha/\sin\theta)$, where α is the angle between the plane "line of sight-magnetic field" and the plane "line of sight-normal to the gap" (supposed to coincide with the axis of the grain).

It is seen from Fig. 2 that FTR model predicts for "antiflakes" the wavelength dependence of the polarization with the maximum at $2ndn_2/\lambda_m \simeq 0.25 - 0.5$ (n = 1.75 or 1.5) the curves both for P_λ/P_{max} and $\frac{\Delta m\lambda - A_V}{E(B-V)}$ being surprisingly similar to those observed in the diffuse IM (cf Fig. 7 in Greenberg, 1978). In addition, the correlation between variations of the polarization maximum λ_m and of extinction curve when changing the thickness of the gaps is of the same sign as that observed from region to region on the sky.

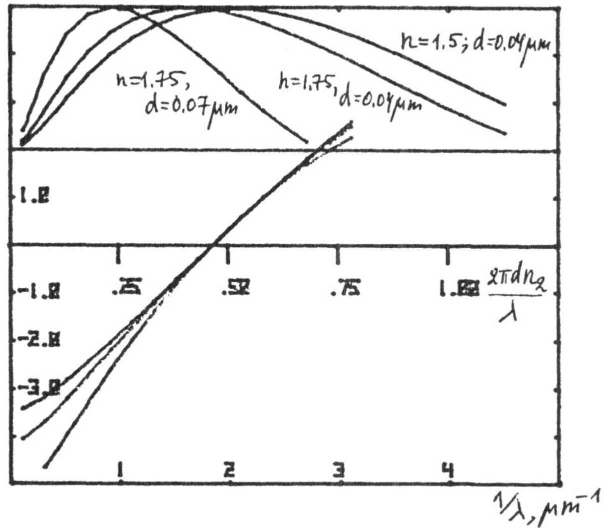

Fig. 2 The wavelength-dependences of the normalized extinction and polarization predicted by the "antiflakes" FTR-model are similar to those observed for the diffuse IM (the cases $n_1 = n_3 = 1.75$ and 1.5 are shown with the gaps d = 0.04-0.07 μm thick for the transverse magnetic field $\gamma_B = 90°$). The predicted correlation of m_λ and p_λ-variations is also of the same sign as the observed one (Greenberg, 1978, p. 197).

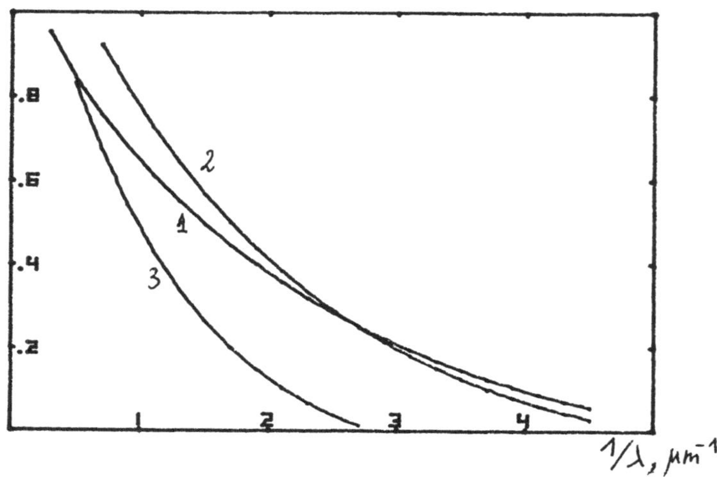

Fig. 3 The ratios $p_\lambda/\Delta m_\lambda$ versus $1/\lambda$ for pure FTR-grains : 1 - n = 1.5, d = 0.04 μm ; 2 - n = 1.75, d = 0.04 μm ; 3 - n = 1.75, d = 0.07 μm.

The FTR ratio $p_\lambda/\Delta m_\lambda$ is very large for $1/\lambda < 2\ \mu m^{-1}$ compared to the average observed value $p_{max}/A_V = 0.03$. The model considered predicts the turnover of the polarization direction by 90° at $\lambda \simeq 0.4\ \lambda_m$ and is different from the widely accepted elongated cylinders' model only by the wavelength dependence of the circular polarization that does not change sign with λ. This can be used in principle for distinguishing the composite grains from single elongated particles within dust-molecular clouds.

With the thickness of ice gaps larger than ∿ 0.05 - 0.07 μm that are consistent with estimates by Greenberg (1982) the polarization maximum will move to wavelengths longer than 1 μm and become narrower which corresponds to the observational data (Wilking 1981). If the core-mantle grains are cemented by ice within dust-molecular clouds the observational evidence for ice to be necessary condition for the linear polarization of embedded infrared sources (Joyce and Simon, 1982) is naturally explained by "antiflakes" in coagulated grains.

REFERENCES

GREENBERG J.M. 1978, Cosmic Dust, Chapter 4, ed. J.A.M.Mc Donnell, J. Wiley § Sons Ltd.

GREENBERG J.M. 1982, in "Submillimeter wave astronomy" ed. J.E. Beckman and J.P. Phillips, Cambridge University Press, P. 261.

HALL E.E. 1902, Phys. Rev. 15, 73.

JOYCE R.R. and SIMON T. 1982, Astrophys. J. 260, 604.

MARTIN P.G. 1972, Mon. Nat. Roy. Astr. Soc. 159, 179.

WILKING B.A. 1981, Astrophys. Y. (see in Ap. J. 1983, 272, 551).

WILLNER S.P. et al. 1982, Astrophys. Y. 253, 174.

THE COEXISTENCE OF SPECTRAL FEATURES OF DUST PARTICLES AND OF IONIZED GAS AS FOUND IN IRAS DATA

M. de MUIZON[1,2] and H.J. HABING[1]

1- Sterrewacht Leiden, 2300 RA Leiden, Nederland
2- Observatoire de Paris, France

ABSTRACT

We present some preliminary data obtained by the Low Resolution Spectrometer onboard IRAS, in the wavelength range 8-22.5 μm. Spectra of about 900 point sources brighter than 20 Jy at 12 μm and than 100 Jy at 25 μm have been examined. Two thirds of the sample could be classified according to the shape of their spectra. Emission lines are seen in the spectra of 118 sources, of which 98 (83%) show the "unidentified" 11.3 μm feature, together with the 7.7 μm and 8.6 μm, assumed to be due to dust particles. Among these latter sources, 90% also show the [NeII]12.8 μm line, and 50% show in addition the [SIII]18.7 μm line.

This suggests that the "unidentified dust features" occur frequently in strong compact infrared sources, and mainly in regions where the dust is associated with ionized gas.

I. DESCRIPTION OF THE IRAS LOW RESOLUTION SPECTROMETER (LRS)

The IRAS* satellite carried a Low Resolution Spectrometer (LRS) as part of its detection system (Fig. 1). Any source observed by the survey instrument was also observed by the LRS, but, during the off-line data reduction, only spectra of sufficiently bright sources were extracted from the data stream. A detailed description of the LRS was given by Wildeman et al. (1983) and we will merely summarize here the main characteristics of the instrument.

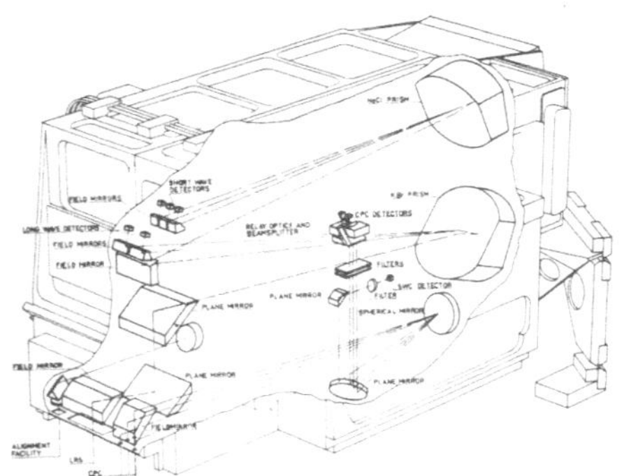

Figure 1

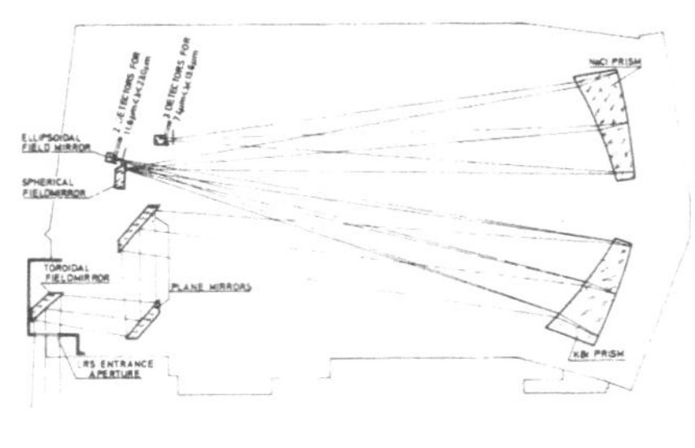

Figure 2

Fig. 2 shows the optical lay-out of the LRS. It is essentially an objective prism. The wavelength range is 8 to 22.5 μm and the LRS aperture is 6' (in scan) x 15' (cross

*The Infrared Astronomical Satellite was developed and is operated by the Netherlands Agency for Aerospace Programs (NIVR), the U.S. National Aeronautics and Space Administration (NASA) and the U.K. Science and Engineering Research Council (SERC).

scan). The wavelength range was divided into two halves which were recorded simultaneously: Band 1 (8-13.5 μm) equipped with 3 detectors Si:Ga, 6'x5' each
Band 2 (11-22.5 μm) equipped with 2 detectors Si:As, 6'x7.5' each.
The long wavelength section consists of a single optical element; it is a reflecting prism with curved optical surfaces combining the functions of collimation, dispersion and re-imaging. For the short-wavelength section, a second similar prism is used in tandem with the first one in order to achieve a better dispersion. In each wavelength band the resolution $\lambda/\Delta\lambda$ varies from 10 at the short wavelength end to about 40 at the long wavelength end. The noise equivalent flux density (NEFD) varies with wavelength between 2 and 4 Jy per spectral resolution element. Because the instrument works as an objective prism it is best suited for point sources. Fig. 3 shows the relative positions of the LRS spectral bands versus the four survey bands centered at 12, 25, 60 and 100 μm.

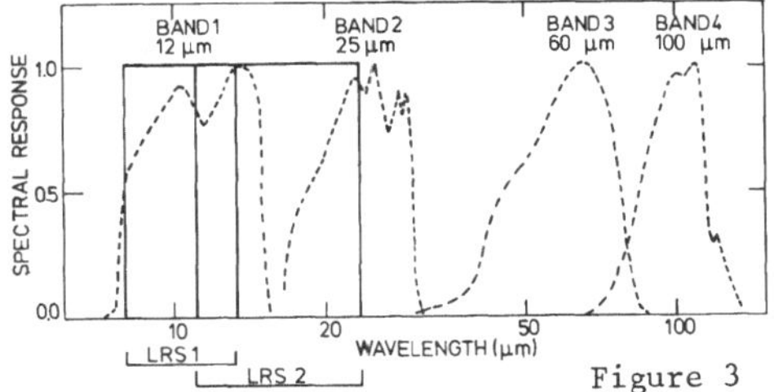

Figure 3

II. OVERVIEW OF THE LRS DATABASE

The LRS database contains to date a set of 171326 spectra, corresponding to about 50000 sources, and formed by extracting from the data stream any source that satisfied the criterium (S/N) survey > 25 both at 12 μm and at 25 μm. This roughly means that the LRS sources are brighter than 2 Jy at these two wavelengths. Practically only 10 to 20 percent of the LRS database will be usable for astrophysical purposes. The remaining spectra either have a signal-to-noise ratio too low, or suffer confusion (in the case of multiple or extended sources); in both cases an astrophysical interpretation will be very difficult if at all possible.

As a first systematic exploration of the LRS database, we have examined spectra of 872 sources, brighter than 25 Jy at 12 μm and than 100 Jy at 25 μm. We have classified these sources depending on the shape of their spectra. The classes as defined and the number of objects found in each of them are given in Table 1.

TABLE 1

Number of sources	Type of object	Characteristics of the spectrum
65	Star	Rayleigh-Jeans tail of a blackbody spectrum Featureless
7	Asteroid	200-300K Blackbody
145	Oxygen-rich star	Broad silicate emission features centered at 10 and 17 μm
44	Oxygen-rich star in a dust envelope	Self absorption in the 10 μm silicate emission feature, and 17 μm emission
100	OH/IR stars, stars or protostars embedded in dust	More or less strong and broad absorption silicate features at 10 and 17 μm
64	Carbon star	11 μm emission feature assigned to Silicon Carbide
118	Compact HII region, Planetary nebula, Reflection nebula, etc.	Emission lines from ionized gas: [SIV]10.5 μm, [NeII]12.8 μm, [NeIII]15.5 μm, [SIII]18.7 μm, etc...And/Or "Unidentified" emission features from dust: 7.7 μm, 8.6 μm, 11.3 μm, etc...

About 38% of the initial sample studied could not readily be classified, mainly because of confusion problems. From now on, we discuss only sources from the last class of Table 1.

III. EXAMPLES OF LRS SPECTRA WITH EMISSION FEATURES AND LINES

1) Comparison with previously published data

Fig. 4a and 4b show some plots of the LRS spectra of the two H II regions S156 and S106, superimposed on data obtained by different groups in different wavelength ranges and compiled by Herter et al. (1982). These spectra show the presence of at least the 11.3 μm feature, and the [NeII]12.8 μm and [SIII]18.7 μm fine-structure lines (of

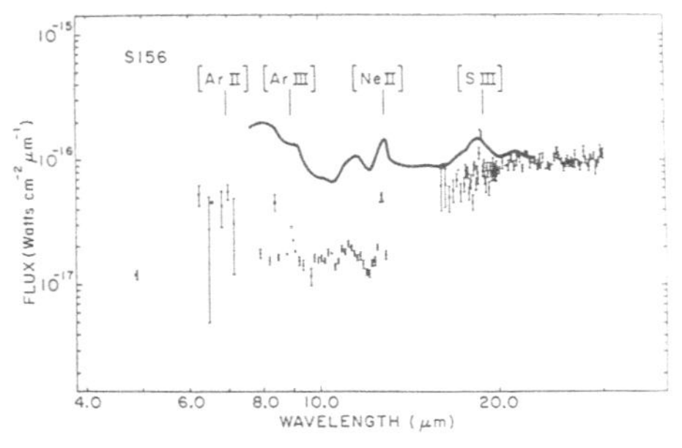

Figure 4a

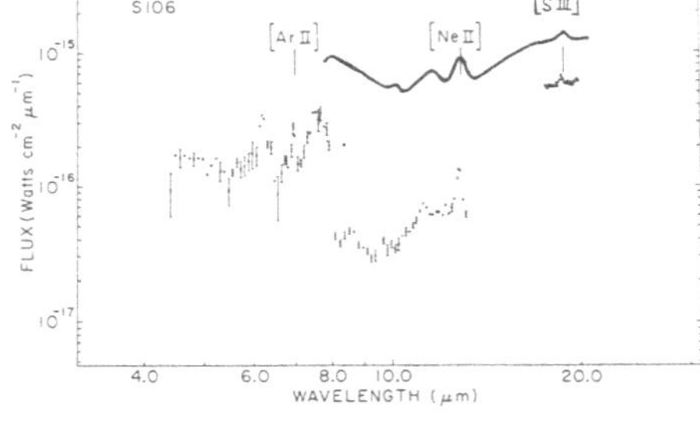

Figure 4b

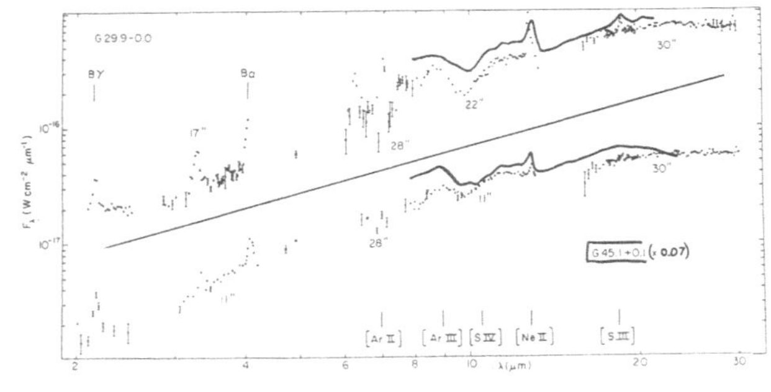

Figure 5

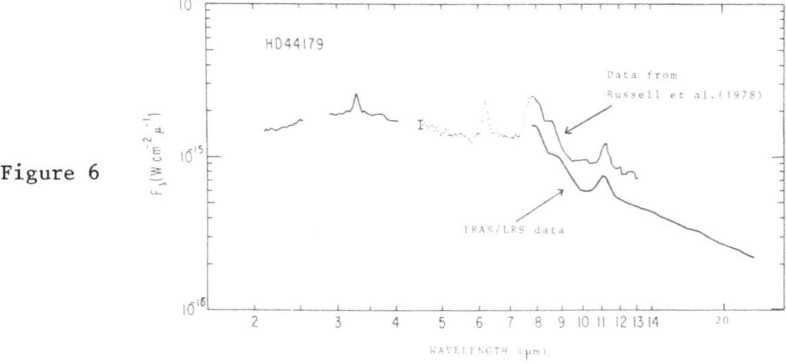

Figure 6

course unresolved by the LRS). The LRS data match reasonably well the shape of the published spectra and contain the same lines. The intensities do not match that well although they do not differ dramatically. But let us keep in mind that the different pieces of the published spectra have not been corrected for beam size effects, when different beam sizes had been employed in different wavelength regions. The LRS 8-22.5 μm spectra have the advantage of having been measured by one instrument, with the same aperture and in similar conditions. Fig. 5 shows the LRS spectra, superimposed on published data compiled by Herter et al. (1981) for the two galactic radio sources G29.9-0.0 and G45.1+0.1. They are characterized by the presence of the [NeII]12.8 μm and [SIII]18.7 μm lines as well as a 10 μm absorption. There is no evidence for the dust emission features. The 2-13 μm spectrum of the reflection nebula called the Red Rectangle and associated with HD44179 is shown on Fig. 6 (Russell et al., 1978). The LRS 8-22.5 μm spectrum is plotted on the same figure; the 8-13 μm spectra agree reasonably well and the longer wavelength part does not show any sign of the presence of gas emission lines. Here is a strong source in which the dust emission features are all observed, but there is apparently no [NeII] or [SIII].

2) Present status of the "unidentified" dust features

The subject has been reviewed by Aitken (1981) and more recently discussed by Allamandola (1984) and Willner (1984). We will recall very briefly here the main properties and possible explanations of these features. A series of six emission features at 3.28, 3.4, 6.2, 7.7, 8.6 and 11.3 μm were first found in the planetary nebula NGC7027, although at different times by different authors (Gillett et al., 1973; Russell et al., 1977). Since then, the features have been found to occur simultaneously and in a variety of objects such as planetary nebulae, H II regions, stellar objects, galaxies and unclassified nebulae. A common point to all the emitting regions is that they have a plentiful supply of UV photons. The features do not break up into lines when observed at higher resolution (Bregman, 1977; Tokunaga and Young, 1980) and therefore they most likely arise from solid grains. Previously, two mechanisms had been proposed to account for these emission features, namely, infrared fluorescence of small molecules (H_2O, CH_4, CO, ...) frozen on grains (Allamandola et al., 1979), and thermal excitation of small grains heated by absorption of UV radiation (Dwek et al., 1980). However, both these explanations meet severe problems (see the review by Allamandola, 1984). Then, Duley and Williams (1981) proposed that the features could arise from small quantities of surface groups bound to chemically active sites on the surfaces of carbon particles. Resonances in aromatic CH groups would be responsible for the 3.3 μm and 11.3 μm features. The explanations seem thus to converge towards the idea of very small grains and aromatic molecules and recently Léger and Puget (1984) have developed this idea much further. On the basis of the observational work by Sellgren (1984), they propose that the emission features arise from extremely small graphite grains, typically 10 Å in diameter. The size of these grains is in fact compatible with complex molecules of unsaturated polycyclic hydrocarbons; one of them, the coronene molecule ($C_{24}H_{12}$) provides a rather good fit to most of the observed features. According to Léger and Puget (1984), the amount of carbon locked up in these molecules would represent up to a few percent of the cosmic abundance of carbon. Moreover, and since the emission features are observed in a variety of objects, these authors conclude that this new population of grains is a general characteristic of the interstellar medium. However, the feature at 3.4 μm, and another feature at 3.5 μm observed only in very few sources, are not explained by their model. All this shows how critical the study of these infrared emission features is for our understanding of the composition of interstellar dust.

3) Correlation of fine-structure lines and "unidentified" features in the LRS spectra

We present in this section the LRS spectra of five infrared sources (Fig. 7). Four of them were already known sources and one is a new IRAS source. Although these data are preliminary and the absolute calibration is still to be improved, the presence of the 11.3 μm emission feature, the [NeII]12.8 μm and the [SIII]18.7 μm emission lines very much degraded by the low resolution, is clear. In addition, on the short wavelength edge of the spectrum, nearly all these objects show what is very likely the long wavelength slope of the 7.7 μm feature, with the 8.6 μm feature appearing as a shoulder. The spectra also show a more or less deep silicate absorption band at around

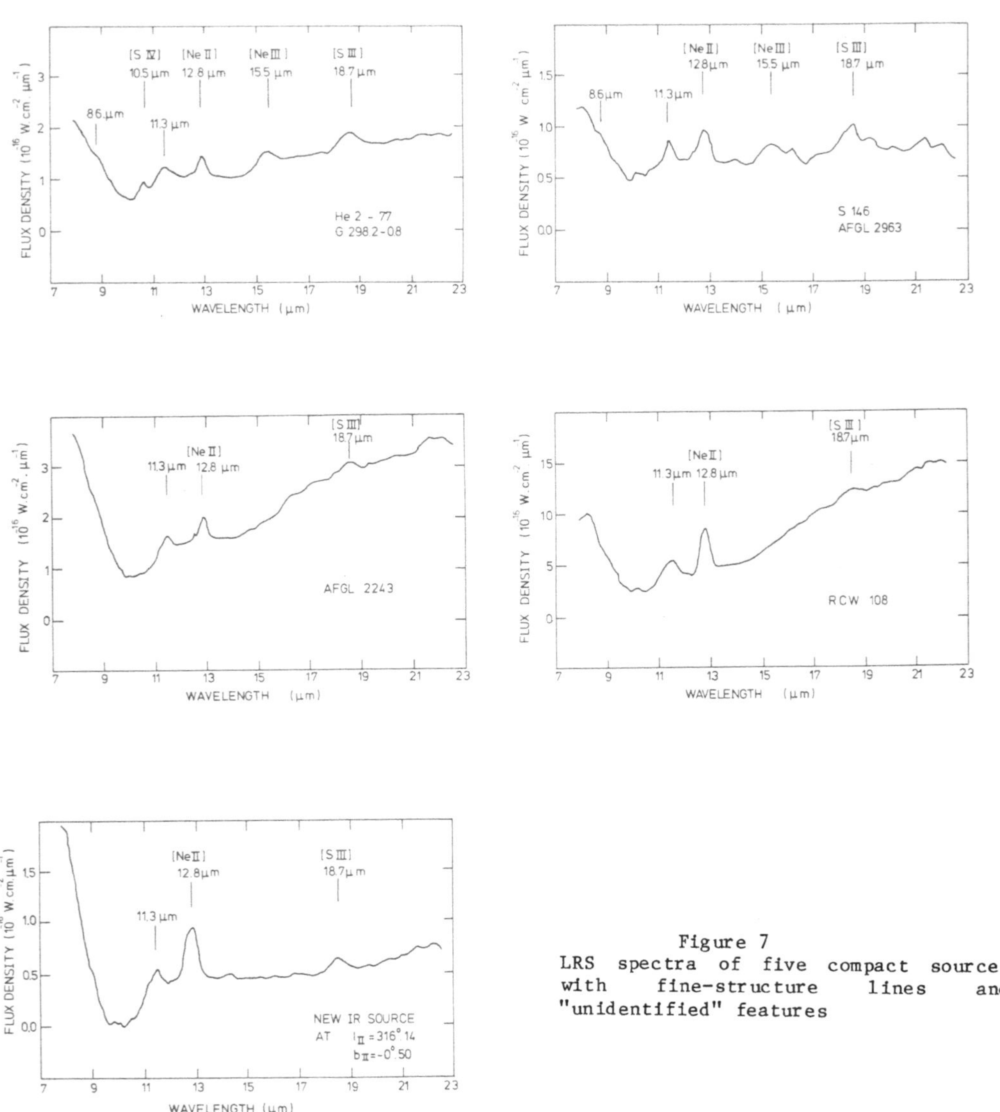

Figure 7
LRS spectra of five compact sources with fine-structure lines and "unidentified" features

10 µm. The four objects previously identified had already been measured in the radio continuum and have 5 GHz fluxes ranging from 1 to 10 Jy. They seem to be heavily obscured compact H II regions with a typical A_v of 10. Preliminary values of the 60 and 100 µm IRAS flux densities show conclusively that the slope of the near to far infrared spectrum is positive.

IV. STATISTICS AND CONCLUSION

The basic sample we have looked at contained 872 sources. Among them 118 sources have spectra with various emission lines and/or features. 98 of them present at least the 11.3 μm feature, and in most cases 7.7 μm and 8.6 μm are quite clear too. In this set of 98 sources, we found the [NeII]12.8 μm line in 90% (88 sources) and both the [NeII]12.8 μm and the [SIII]18.7 μm lines in 50% (48 sources).

The following conclusions can immediately be drawn from these numbers: i) the "unidentified" feaures are very common in strong compact sources of infrared emission, ii) in these objects the features seem to occur mainly when the dust is associated with ionized gas.

It is not clear however whether the association of ionized gas with the grains producing the features still exists in the diffuse medium. Indeed, those grains are suspected to be also responsible for the high colour temperatures measured in some extended nebulae, in places where there is no evidence of the presence of ionized gas (A. Leene and C.A. Beichman, this volume).

Nevertheless, the LRS database gives us a rather large non biased sample to investigate compact sources showing the "unidentified" features from hot dust, or the fine-structure lines from ionized gas, or a combination of both. Further studies of all these features and lines, as well as follow-up observations at higher spectral and spatial resolution should help us to set constraints on the physical properties of the objects in question (excitation conditions, temperatures, abundances, densities, etc.) and to define more precisely which conditions give rise to particular features.

References

Aitken, D.K., 1981, in Infrared Astronomy, IAU Symposium No. 96, eds. C.G. Wynn-Williams and D.P. Cruikshank, Reidel, Holland, p.207.

Allamandola, L.J., 1984, in Galactic and Extragalactic Infrared Spectroscopy, XVIth ESLAB Symposium, eds. M.F. Kessler and J.P. Phillips, D. Reidel, ASSL Vol. 108, p.5.

Allamandola, L.J., Greenberg, J.M., and Norman, C.A., 1979, Astron. Astrophys. 77, 66.

Bregman, J.D., 1977, P.A.S.P. 89, 335.

Duley, W.W. and Williams, D.A., 1981, M.N.R.A.S. 196, 269.

Dwek, E., Sellgren, K., Soifer, B.T., and Werner, M.W., 1980, Astrophys. J. 238, 140.

Gillett, F.C., Forrest, W.J., and Merrill, K.M., 1973, Astrophys. J. 183, 8.

Herter, T., Helfer, H.L., Pipher, J.L., Briotta, D.A., Forrest, W.J., Houck, J.R., Rudy, R.J., and Willner, S.P., 1982, Astrophys. J. 262, 153.

Herter, T., Helfer, H.L., Pipher, J.L., Forrest, W.J., McCarthy, J., Houck, J.R., Willner, S.P., Puetter, R.C., Rudy, R.J., and Soifer, B.T., 1981, Astrophys. J. 250, 186.

Léger, A., and Puget, J.L., 1984, Astron. Astrophys. 137, L5.

Russell, R.W., Soifer, B.T., and Willner, S.P., 1977, Astrophys. J. 217, L149.

Russell, R.W., Soifer, B.T., and Willner, S.P., 1978, Astrophys. J. 220, 568.

Sellgren, K., 1984, Astrophys. J. 277, 623.

Tokunaga, A.T., and Young, E.T., 1980, Astrophys. J. 237, L93.

Wildeman, K.J., Beintema,, D.A., and Wesselius, P.R., 1983, J. British. Interplanet. Soc., 36, 21.

Willner, S.P., 1984, in Galactic and Extragalactic Infrared Spectroscopy, XVIth ESLAB Symposium, eds. M.F. Kessler and J.P. Phillips, D. Reidel, ASSL Vol. 108, p. 37.

CO OBSERVATIONS OF HIGH VELOCITY GAS AROUND S187

Casoli F., Combes F., Gérin M.
Ecole Normale Supérieure, Paris V°
and Observatoire de Meudon, 92190 Meudon, France

I- Introduction and observations:

Sharpless 187 is an HII region located in the Orion arm, in the second galactic quadrant. Two sources have been observed at radio wavelengths by Israël (1977), one compact source S187A, whose relation to the Sharpless region is not certain (Israël, priv. com.) and a more extended (2.4') region S187B. Far Infrared observations by Sargent et al. (1981) reveal two sources, suggesting emission from dust heated by two embedded objects, one identified with the compact S187A, the other being the exciting star (B0) of the evolved nebula. We report here large-scale observations of CO and 13CO together with higher resolution CO (2-1) observations around S187. The distance of the source, in the Orion arm, is taken to be 800 pc.

The large-scale 13CO observations were part of a mini-survey of the Orion arm in the 2nd quadrant with the 2.5m millimetric antenna in Bordeaux (Casoli et al., 1984). The region was fully sampled, with the beam of 4.4' (=1pc), over 1.5 square degrees. The velocity resolution was 0.27km/s, and we integrated to obtain a noise level of 0.1K all over the map.

The CO (2-1) observations were done with the 5m antenna of the MWO at Mac Donald Observatory, Texas. The spatial and velocity resolutions were respectively 1.4' (=0.32pc) and 0.32Km/s.

II- Collision of two molecular clouds?:

The large-scale 13CO observations (fig 1) reveal two velocity components, which move one relative to the other systematically until the central source, where there remains only one component. In the center the integrated 13CO emission is higher than the sum of the 2 components elsewhere in the map.

The CO spectra in the same region are much wider and do not present the same characteristics: in particular, the more positive velocity component never disappears completely, even in the center, where the ratio $12T_A$ / $13T_A$ reaches $\geqslant$ 50 (optically thin component).

This peculiar velocity structure can be interpreted by the collision of two molecular clouds, the front of which gives rise to gas compression and star formation. The H_2 mass derived for the two clouds from the 13CO emission over a size of 15pc is 2 10^4 Mo.

Figure 1: 13CO spectra around S187 (Bordeaux telescope).

20'

40'

δP

60'

20

10'

δb

0

T_a^*

2 K

-20. 0.

V (Km/s)

III-High velocity flow:

The CO(2-1) emission (fig2-3) reveal high velocity wings at low antenna temperature (less than 1K). We have integrated this emission between -26.7 and -18.9 Km/s (blue wing) and -8.7, -0.9Km/s (red wing): The corresponding isophotes are displayed in fig. 4. The blue and red maps have not the same center, there is anisotropy of the emission. The infrared source and compact HII region are located between the blue and red map centers. The geometry of the source is very similar to NGC 2071 (Bally, 1982). The mean direction of the flows is inclined by only a small angle to the line of sight. Bally & Lada (1983) have observed in CO(1-0) the central position of S187 and also noticed significant high-velocity wings.

The red wing integrated emission is more intense and more extended than the blue wing. This can be explained by radiation transfer effects: the regions of the flows near the central source are hotter and absorption by cooler gas occurs only for the approaching flow.

Physical conditions: The 12CO / 13CO antenna temperature ratio is 25 in the wings: this corresponds to an optical thickness of 1 for CO and .025 for 13CO (the abundance ratio is taken to be 40). The total H_2 mass can then be estimated at 10M⊙ (with an uncertainty of a factor 2). The excitation temperature would be 8K, if there were no beam dilution. But the CO(2-1) / CO(1-0) antenna temperature ratio is of the order of 1, and T_{ex} must be much higher. There is a surface filling factor of about 15% for the clumps in the flow.

The maximum velocity of the molecular flow is V=13Km/s, and the extension is R=2pc (geometrical mean in the two projected directions). The dynamical time R/V is then 1.5 10^5 years. With the estimated mass of 10M⊙, the corresponding average volumic density in the flow is n = 60 at cm^{-3}. These values correspond to a rather old event, relative to the other detected flows, and the density follows the $n \sim R^{-2}$ relation of Bally & Lada (1983).

References:

Bally J. (1982) Ap. J. 261, 558

Bally J., Lada C.J. (1983) Ap. J. 265, 824

Casoli F., Combes F., Gérin M. (1984) A.A. 133, 99

Israël F.P., (1977) A.A. 61, 377

Sargent A.L, van Duinen R.J., Nordh H.L., Aalders J.W.G. (1981) A.A. 94, 377

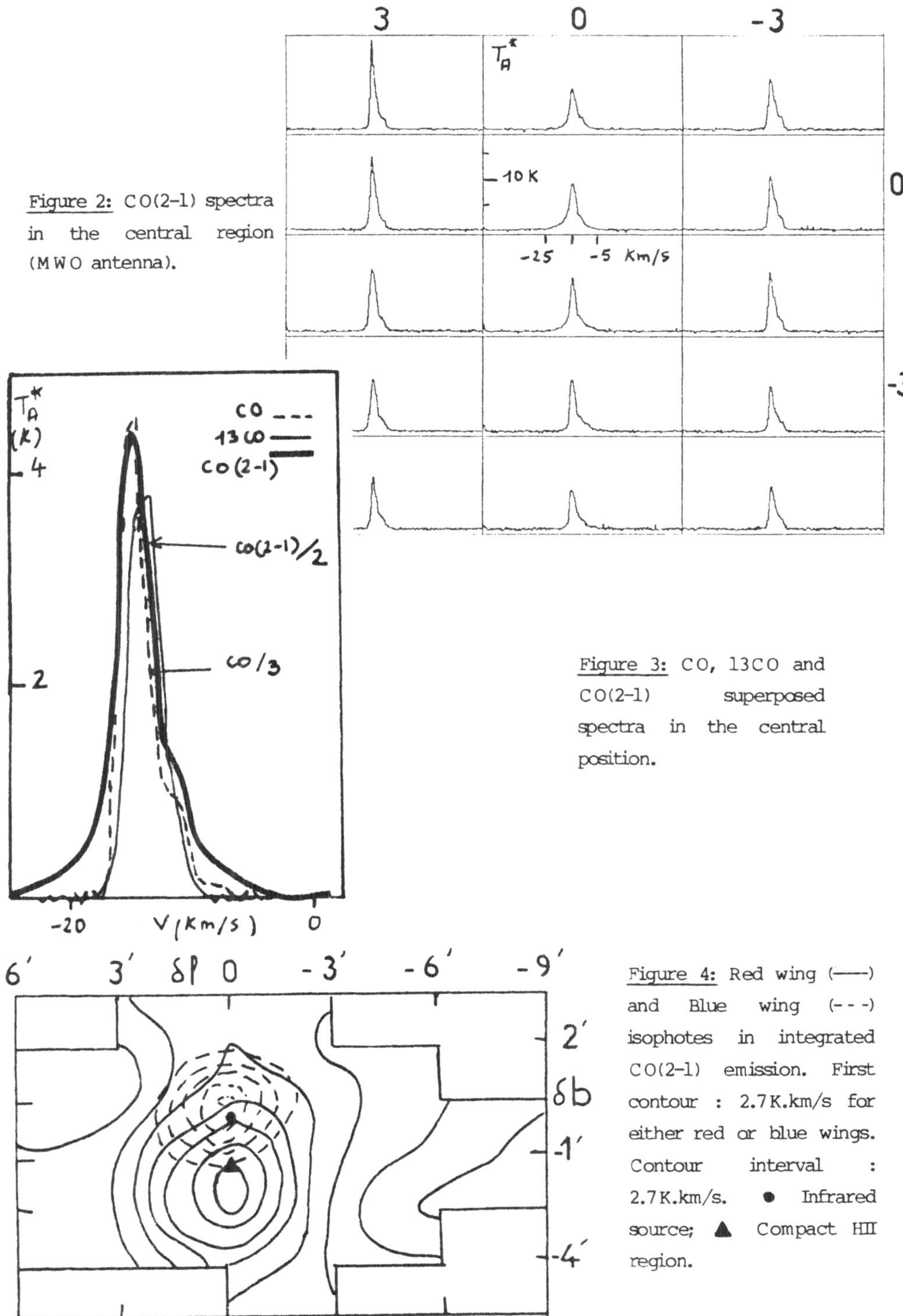

Figure 2: CO(2-1) spectra in the central region (MWO antenna).

Figure 3: CO, 13CO and CO(2-1) superposed spectra in the central position.

Figure 4: Red wing (——) and Blue wing (- - -) isophotes in integrated CO(2-1) emission. First contour : 2.7 K.km/s for either red or blue wings. Contour interval : 2.7 K.km/s. ● Infrared source; ▲ Compact HII region.

CO AND ^{13}CO OBSERVATIONS OF HI ABSORPTION CLOUDS

D. Despois, A. Baudry
Observatoire de Bordeaux
33270 Floirac, France

Most HI absorption clouds which have been searched for carbon monoxide emission are local clouds with no heavy obscuration (e.g. Kazès and Crovisier, 1981). However, comparison of the atomic and molecular hydrogen cloud components is now possible throughout the Galaxy because sensitive interferometers produce reliable HI absorption spectra even close to the galactic plane. This is of great significance to studying the spatial and chemical relationship of HI with H2 in interstellar clouds.

We have used the Bordeaux 2.5 m telescope (Baudry *et al.*, 1981) to search for J = 1-0, CO and ^{13}CO emission toward 33 continuum sources selected in the first, second and third galactic quadrants at $|b| < 5°$ from the VLA HI survey of Dickey *et al.* (1983). (Two other sources observed by Crovisier and Dickey, 1983, have been added to this sample.) The sensitivity was uniform : in one 100 kHz channel, σ = 0.4 and 0.25 K in CO and ^{13}CO respectively. At least one CO or ^{13}CO feature is detected in 70 % or 26 % of the selected HI directions while less than 20 % of the total number of HI features (217) have a carbon monoxide counterpart.

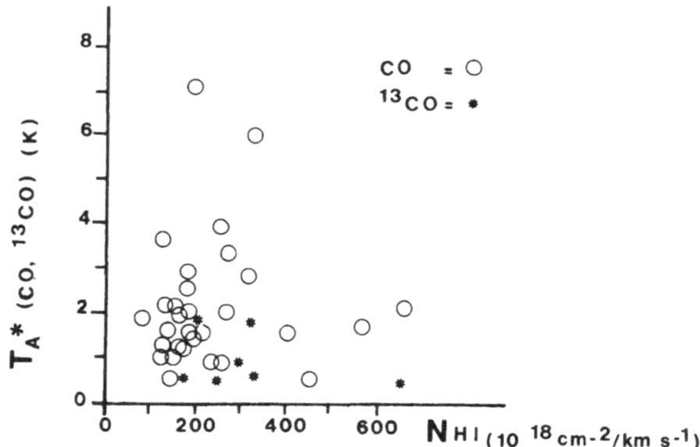

Figure 1. Plot of CO and ^{13}CO intensities against HI column densities per unit velocity.

The histogram of velocity differences, DV, between a CO feature and the HI feature closest in velocity shows that nearly 70 % of the

HI-CO associations have $|DV| < 1.6$ km/s, a number much smaller than the difference between two contiguous HI features (which typically is ~ 8 km/s). Therefore each CO cloud is physically associated with an HI component. However there is no clear correlation of CO and ^{13}CO intensities with T_{spin}, τ_{HI} or (Fig. 1) with $N_{HI}/\Delta V$ although the probability of finding a CO-HI association is shown to increase with τ_{HI} (Figure 2). This probability reaches 65 % in the first galactic quadrant when $\tau_{HI} > 1.6$.

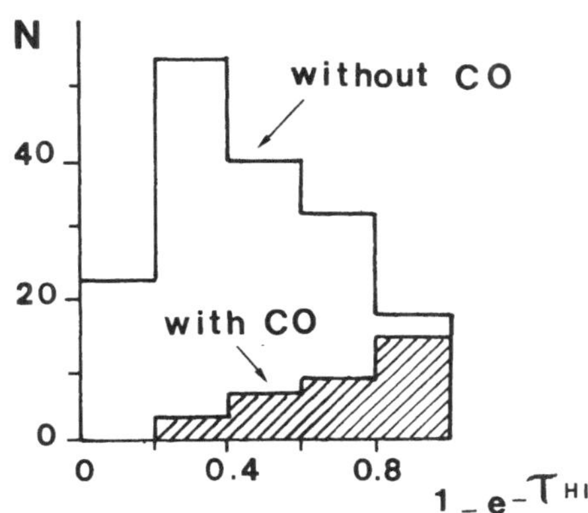

Figure 2. Histogram of HI features with or without CO emission versus $1 - e^{-\tau_{HI}}$.

The galactic distributions of CO and HI have been compared to further investigate the relationship between both species. Z being the height above the galactic plane, we derive : for CO, $<|Z|> = 33$ or 46 pc and $\sigma_{CO} = 49$ or 66 pc ; for HI, $<|Z|> = 59$ or 88 pc and $\sigma_{HI} = 86$ or 136 pc. The two numbers given for each parameter correspond to two different cases where all features in the first quadrant are placed at the near kinematic distance or are split up into near and far distance features. These results obtained without any a priori assumption on the HI and CO distributions show that the HI absorption clouds, although closer to the galactic plane than the HI intercloud medium (cf. Crovisier, 1978, for a review), do not coincide with the distribution of molecular clouds. Finally the following picture is suggested : high opacity HI features lying close to the galactic plane tend to show higher CO probability detection while low opacity HI features, with broader Z distribution, tend to have no molecular counterpart.

Assuming $T_{spin} = T_{ex}(CO)$ and $N_{H2}/\int T_A^*(CO)\,dv = 2\ 10^{20}\ cm^{-2}/(K\ km/s)$ we derive $N_{HI}/N_{H2} \simeq 0.02 - 0.2$ for a set of 21 HI-H2 associations. The hypothesis $T_{spin} = T_{ex}(CO)$ and CO dilution tend to underestimate the HI and H2 column densities respectively.

A detailed account of this work is given by Despois and Baudry (1984). In addition we have observed with the MWO telescope (University of Texas) the J = 2-1 line of CO and ^{13}CO toward several sources selected for their rich atomic and molecular line spectra. Further observations are scheduled on the Onsala 20 m dish. The main purpose of these new observations is to estimate the effects of beam dilution and to improve the comparison of the atomic and molecular hydrogen contents of interstellar clouds.

REFERENCES

Baudry, A., Cernicharo, J., Pérault, M., de La Noë, J., Despois, D. : 1981, Astron. Astrophys. **104**, 101
Crovisier, J. : 1978, Astron. Astrophys. **70**, 43
Crovisier, J., Dickey, J.M. : 1983, Astron. Astrophys. **122**, 282
Despois, D., Baudry, A. : 1984, Astron. Astrophys., accepted for publication
Dickey, J.M., Kulkarni, S.R., van Gorkom, J.H., Heiles, C.E. : 1983, Astrophys. J. Suppl. **53**, 591
Kazès, I., Crovisier, J. : 1981, Astron. Astrophys. **101**, 401

PART V

STELLAR CONTENT OF NEARBY MOLECULAR CLOUDS

THE STELLAR CONTENT OF NEARBY CLOUDS - T TAURI STARS

Claude Bertout
Laboratoire d'Astrophysique théorique du Collège de France
and Institut d'Astrophysique
98bis, boulevard Arago
F 75014 Paris, France

Summary

After a brief summary of the properties of T Tauri stars, we review current mass-loss rate determinations and line profile computations.

I. Introduction

Although there is little observational evidence so far that young low-mass stars influence the evolution of the cloud's bulk, several mechanisms have been proposed which depend upon low-mass pre-main-sequence stars to regulate star formation in molecular clouds. For example, Norman and Silk (1980) proposed that intersecting bubbles caused by stellar winds from T Tauri stars create dense gas clumps and thereby trigger protostellar collapse. However, the magnitude of mass-loss rates in T Tauri winds (see below) and the number of T Tauri stars present in clouds are probably both too low for this scenario to work. Another mechanism proposed by Silk and Norman (1983) calls for the X-ray flux of T Tauri stars to regulate the star formation process by controlling the cloud's ionization degree. This attractive possibility is compatible with known properties of T Tauri stars.

Even if T Tauri stars were not as intriguing as they are, possible large-scale interactions between them and the molecular clouds from which they are born would be reason enough to devote part of this conference to them. The next section will thus review the main properties of T Tauri stars, and their mass-outflows will be discussed in Section III. The last section contains a progress report on current work on T Tauri stars' line profiles, which still remain puzzling in many respects.

II. The Class of T Tauri Stars

First defined by Joy in 1945, the classification criteria of T Tauri stars were further refined by Herbig in 1962. Arguments in favor of an interpretation of T Tauri stars as low-mass, pre-main sequence objects have been presented by Ambartsumian (1947), Walker (1956), Zappala (1962), Herbig (1977), and Jones and Herbig (1979). Main catalogs of T Tauri stars and related objects were compiled by Herbig and Rao (1972), Cohen and Kuhi (1979) and Appenzeller et al. (1983); and recent comprehensive reviews are those of Bertout (1984), Cohen (1984) and Edwards (1985). Below, I summarize T Tauri star's properties and briefly discuss the most recent results.

Optical Spectrum

Optical properties of T Tauri stars are reviewed by Herbig (1962) and Cohen and Kuhi (1979). The photospheric late-type spectrum is often "veiled" (i.e., the absorption lines are filled-up to various degrees) by what now appears to be chromospheric emission. In "extreme" T Tauri stars, this effect makes it impossible to discern the photospheric spectrum but these represent less than 10% of the T Tauri population.

Balmer lines

Often complex and highly variable (Bertout et al. 1982, Mundt and Hartmann 1983, Mundt 1984), they have been interpreted in terms of isotropic outflow by Kuhi (1964). However, a subclass of T Tauri stars, the YY Orionis stars, show inverse P Cygni profiles at their Balmer lines (Walker 1972, Appenzeller 1977). A few stars among the most active show line profile variations from P Cygni to inverse P Cygni on a time scale of days (e.g. Krautter and Bastian 1980). Shapes of Balmer line profiles will be discussed in more detail in Section III.

NaD lines

Spectra of extremely high resolution of the NaD doublet, obtained by Mundt (1984), allowed him to discover narrow blue-displaced absorption components in the NaD lines of a number of stars. He discussed two possibilities for the origin of these absorptions --the cool outer parts of the stelar wind, or wind-driven shells-- and concludes that none of them is satisfactory. On one hand, the radial velocities of the sharp absorption do not seem to be good indicators of the wind's end-velocity; and on the other hand, the observed time variations of the absorption component velocities are not consistent with that expected for wind-driven shells.

Forbidden lines

Appenzeller et al. (1984) study the profiles of various forbidden lines in the spectra of a number of T Tauri stars and conclude that they are probably formed in anisotropic outflows partially occulted by dusty disks surrounding the stars.

Infrared Properties

Mendoza (1966) discovered that T Tauri stars show large infrared excesses while a large body of observations in the near-infrared is presented by Rydgren et al. (1976). Pionniering observations in the far-infrared are those of Harvey et al. (1976). IRAS data on T Tauri stars are discussed elsewhere in this volume.

Polarization

Bastien (1981, 1982a,b) has conducted a large study which shows that polarization in the T Tau class is probably due to the presence of dust anisotropically distributed around the stars. Polarization data relevant to the topic of collimated outflows are discussed by Mundt in this volume.

Ultraviolet Observations

Spectroscopy using the IUE satellite revealed that UV lines of ionized metals were much stronger than in the Sun or even than in active late-type stars such as the RS CVn systems (e.g. Appenzeller et al. 1980; Cram et al. 1980; Jordan et al. 1982). Simultaneous optical and ultraviolet spectroscopy allowed Calvet et al. (1984) to conclude that, similarly to Hα, the MgII h and k lines are formed in a more extended region than the CaII H and K lines.

Properties at X-ray Wavelengths

While a number of individual stars have been observed once or twice at X-ray wavelengths (Ku and Chanan 1979; Gahm 1981; Feigelson and DeCampli 1981; Walter and Kuhi 1981), only the Rho Ophiochi region has been monitored repeatedly (Montmerle et al. 1983); and it demonstrates considerable flare activity. The level of X-ray emission is comparable to that of other late-type active stars. Bouvier et al. (this volume) demonstrate that the relationship between X-ray flux and rotational velocity is the same for T Tauri stars as for other late-type dwarfs and active systems, which hints at the presence of coronae around T Tauri stars.

Radio Data

Recent continuum VLA data are presented and discussed by Montmerle and Feigelson in this volume. Earlier continuum data are discussed in the next section, as are observations of high-velocity molecular gas in the vicinity of T Tauri stars.

III. Mass-loss in T Tauri Stars

1. Theoretical Expectations

DeCampli (1981) must be credited for having been first to emphasize the problems of strong T Tauri winds. Wind-driving mechanisms based on radiative pressure, either directly as in early-type stars or indirectly through grain acceleration can only account for mass-loss rates of less than 10^{-9} $M_{\odot}/yr$ in T Tauri stars because of their low luminosities. Because of the moderate X-ray luminosities of T Tauri stars, the same upper limit for M is found for winds driven by Parker-type coronal expansion, and their relatively low rotational velocities (Vogel and Kuhi 1981; Bouvier et al., this volume) are inconsistent with models which assume that the wind is driven by rotation.

DeCampli proposed that Alfvén-waves were driving the flow, and estimated that a maximum mass-loss rate of about $3\ 10^{-8}$ $M_{\odot}/yr$ could be obtained from this mechanism. Quantitative models of Alfvén-wave driven winds were computed by DeCampli (1981), Hartmann, Edwards and Avrett (1982) and Lago (1979, 1984).

Holzer, Flå and Leer (1983) however concluded from a detailed study of the role of wave-damplíng that Alfvén-waves would be unlikely to drive the massive, low-velocity winds observed in red giants. Whether their criticism is invalid in the case of T Tauri stars, which have presumably higher wind terminal velocities, remains to be tested. Holzer et al. also caution about using a constant damping length in the computations of wave-driven winds, since the radial runs of physical variables, particularly the temperature, turn out to be very different when the variation of the damping length with radius is computed self-consistently. The models of Hartmann et al. (1982), which were computed under the assumption of a constant damping length (equal to the stellar radius) must therefore be approached with some caution. Obviously, a detailed discussion of the Alfvén-wave damping lenghts in T Tauri envelopes is needed if the validity of this wind-driving mechanism is to be confirmed.

2. Observational Evidence

Traditionally, mass-loss rates of T Tauri stars were derived from the study of their hydrogen line profiles. Herbig (1962) and Kuhi (1964) found relatively high mass-loss rates for several stars (e.g., $\dot{M}=10^{-6}$ $M_\odot$/yr for V1331 Cyg and 2 10^{-7} $M_\odot$/yr for T Tauri). It has since then become apparent that deriving mass-loss rates from a study of line profiles alone is a particularly difficult task, and these early results have been challenged (cf. DeCampli 1981). In the following, I shall review both the new methods used to derive mass-loss rates from T Tauri stars and the current state of affairs in line profile modeling.

a. Radio continuum

For some time in 1981-82 it looked as if the problems of determining mass-loss rate in T Tauri stars would finally have an easy and unambiguous solution. Some of the stars were being detected in the radio continuum, which would allow accurate determination of the mass-loss rate provided that the emission is thermal, that the emitting envelope is spherically symmetric, and that the end-velocity of the mass-flow is known. With these assumptions, Bertout and Thum (1982) derived a mass-loss rate of about 4 10^{-8} $M_\odot$/yr for DG Tauri. A main source of incertainty when determining mass-loss rates from radio thermal continuum data is the required assumption about the wind's end-velocity. One usually assumes that the Hα full-width at half-maximum is a measure of the end velocity, but several other indicators have been proposed. If, on one hand, the velocity of the blue edge of forbidden lines is a good indicator of the wind's terminal speed as was proposed by Appenzeller et al. (1984), smaller values of final velocities result. If, on the other hand, the speed of Herbig-Haro objects is the relevant indicator (Mundt 1984), then larger values are expected. Since we understand the physics of forbidden lines much better than that of Herbig-Haro objects, it seems safe to conclude that the FWHM of Hα probably yields an upper limit of the wind's end-velocity, and hence of the mass-loss rate.

The first VLA results (Cohen, Bieging and Schwartz 1982) brought in some surprises, since radio emission from the T Tauri region turned out to originate from the infrared companion of T Tauri (de Vegt 1982; Hanson et al 1983). Additional VLA observations of that region (Simon et al. 1983) allowed the radio emission of the optical component of T Tauri to be resolved, which yields a mass-loss rate of about 4 10^{-8} $M_\odot$/yr for this star as long as the flow is isotropic. This upper limit is much smaller than the value derived by Kuhi (1964) from the modelling of optical lines.

As more stars are observed in the radio range and as more information on their variations becomes available, the picture becomes however increasingly complicated. Radio emission of objects such as DG Tauri seems fairly constant and extended so that for this star the assumption of thermal free-free radiation is probably valid. But to complicate matters, Bieging, Cohen and Schwartz (1984) find that radio emission of DG Tau is not isotropic, but rather extends in the same direction as the optical nebula seen by Mundt and Fried (1983). Although we do not expect this anisotropy to change the variation-law of the radio flux with frequency (Schmid-Burgk, 1982), evidently only an upper limit to the mass-loss rate can be derived from the radio continuum flux emitted in non-spherical envelopes.

For objects such as V410 Tauri no statement about the mass-loss rate can be made from the radio flux, since observed flux variations by a factor 10 or more on an unknown time-scale (Bertout 1984; Cohen, private communication) suggest non-thermal events. Since V410 Tauri is also the fastest known rotator of the T Tauri class, it seems natural to hypothetize that its radio emission is caused by magnetic activity. However, this may be too naive a representation as Montmerle and Feigelson show for the Rho Ophiochi stars in this volume. The only conclusion that can be safely drawn at this point is therefore that very few of the T Tauri stars detected in the radio range have thermal radio spectra.

b. High-Velocity Molecular Outflows

Observations of large-scale, often bipolar, high velocity molecular outflows provided the first evidence of anisotropic mass-loss in the vicinity of T Tauri stars (Bally and Lada, 1982; Edwards and Snell, 1982; Calvet, Canto and Rodriguez, 1984). Because these outflows are most probably driven by the momentum of a stellar wind, one can derive properties of the stellar wind from the study of the molecular outflow as has been done by the authors cited above; for example, Edwards and Snell found mass-loss rates greater than the theoretical limits discussed above (e.g., $\dot{M}=2\ 10^{-7}\ M_{\odot}/yr$ for T Tauri). The derivation of mass-loss rates from the properties of the molecular flows requires, however, a number of assumptions that can affect the quantitative results by large factors. Neither Edwards and Snell nor Bally and Lada take projection effects in the velocity field into account, but assume instead that the flow has a single characteristic velocity. This assumption is responsible for most of the discrepancy between Edwards and Snell's results and the more moderate mass-loss rates derived by Calvet, Canto and Rodriguez.

We find that the best estimate for the mass-loss rate of T Tauri itself as derived from its associated molecular outflow is 1-2 10^{-8} $M_{\odot}$/yr, with an uncertainty factor of about 4 (Bertout and Cabrit in preparation). This value is compatible with the upper limit of the mass-loss rate derived from radio data. Detailed molecular line profile computations are however necessary in order to get better estimates of the molecular flow properties and thus of the stellar mass-loss rate.

c. Line profiles

Ulrich and Knapp's (1979) atlas of observed profiles as well as high-resolution profiles presented by Mundt (1984) both confirm Kuhi's (1978) results that presented statistics of Hα line profile shapes in the T Tauri class. There are three basic Hα line profile shapes with a number of possible variations. Classical P Cygni profiles, Type I in Beals' 1950 classification, have been observed for a few stars (Figure 1); but P Cygni profiles of Type III can be said to be more characteristic of the T Tauri class, since they are exhibited by the Hα line of more than 60% of the stars (Figure 2). Type III profiles have two emission peaks separated by a usually blue-displaced minimum. The velocity of the minimum is typically 80km/sec, but can range from 0 to more than 200km/sec in different stars (Herbig 1977). Another large fraction of T Tauri stars (about 30%) has symmetrical profiles with little structure (Figure 3).

High mass-loss from T Tauri stars was first suggested by profile computations of various spectral lines, particularly of Hα . In recent years, it has become increasingly clear that computed profiles are extremely sensitive both to the details of the adopted envelope models (cf. Wagenblast, Bertout and Bastian 1983) and to the various approximations used in the radiative transfer treatment (Calvet, Basri and Kuhi 1984). Modelling one or even a few line profiles is thus not sufficient for an unambiguous determination of the physical properties of the emitting atoms. The remedy is to use a battery of diagnostic tests, e.g., to model several lines with different excitation conditions or to combine line profile computations with other mass-loss derivation methods. The latter approach is possible only for a few well-studied stars, such as T Tauri itself or DG Tauri. Progress is rapidly being made in different aspects of the radiative transfer problem, so that the determination of mass-loss rates through multiple line computations will become a common procedure in the near future; but for the time being, we must conclude that no reliable mass-loss rates have been derived on the basis of line profile modelling yet.

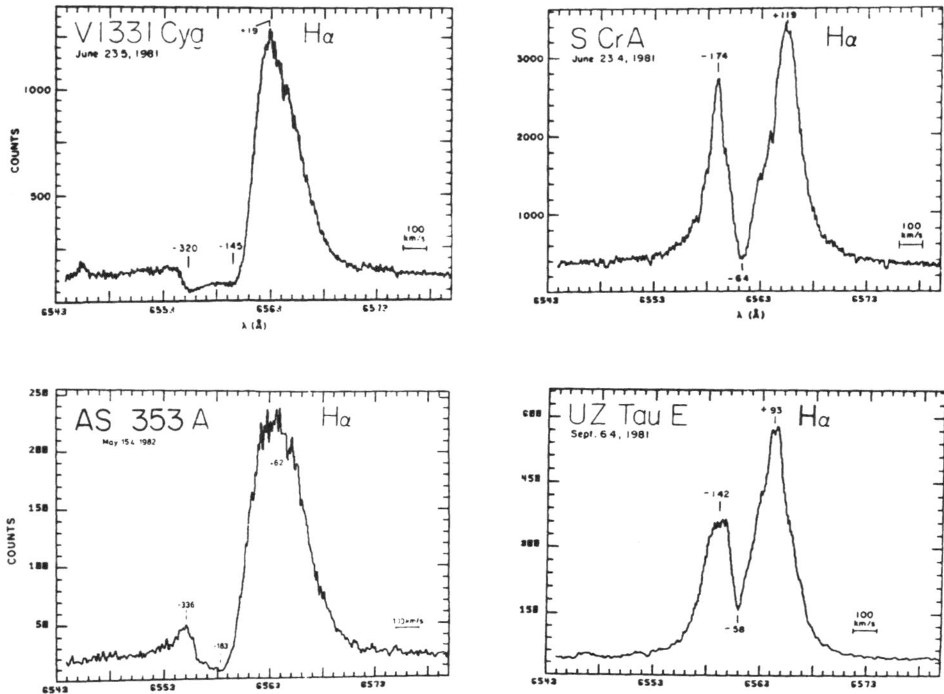

Fig.1 : Observed Hα profiles in V1331 Cyg and AS353 (from Mundt 1984)

Fig.2 : Observed Hα profiles in SCrA and UZ Tau E (from Mundt 1984)

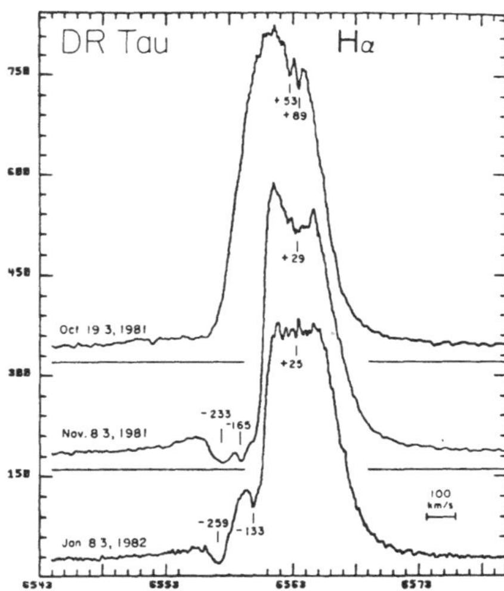

Fig.3 : Observed Hα profiles in DR Tauri (from Mundt 1984)

3. Conclusions

The above discussion of results obtained by several groups show that there is no evidence at this point for any mass-loss rate greater than about 10^{-8} $M_{\odot}$/yr among the T Tauri stars for which apparently thermal radio spectra have been observed, or among those with associated molecular outflows. We should emphasize here that such data exist only for a few of the most remarkable T Tauri stars. In particular, there are no mass-loss rate estimates for the modest, canonical K7 stars which best represent the class. Whether mass-loss mechanisms other than Parker-type coronal expansion need to be considered for the average T Tauri star thus remains to be demonstrated.

IV. Current Line Profile Computations for T Tauri Stars

Several complementary approaches are currently being taken by workers trying to model line profiles of T Tauri stars. The Berkeley group advocates the semi-empirical approach that has been so successful with the Sun. It puts emphasis on resonance lines of various elements, which are supposed to be easier to model since a two-level atom offers a good approximation for such a transition. This simplicity is only illusory, however, since partial redistribution effects must be taken into account when modeling the wings of these lines. Earlier efforts focussed on the chromosphere while neglecting velocity fields (Calvet, Basri and Kuhi 1984), but formation of resonance lines in the presence of velocity fields is now under investigation (Basri, Bertout and Calvet 1984).

Since we don't know anything about outflow geometry, except that it is highly anisotropic, another needed approach is to find out the envelope geometry and velocity field that would most probably lead to the profiles exhibited by the resonance and hydrogen lines of T Tauri stars. This can at least be done without considering all the details of line formation, for one can assume hat the line widening is largely due to organized velocity fields. To illustrate this approach, I show below line profiles formed in biconical geometry (Figure 4) which I assume to simulate (in a first approximation) the extension down to the close stellar environment of the jets found by Mundt and Fried (1984) in the environment of several T Tauri stars. Since evidence for bipolar outflows is on a spatial scale of about 100 AU or more, the question of whether the optical line profiles could be formed in bipolar outflows on smaller spatial scales (about 10 to

100 stellar radii) is indeed an interesting one. Note however that if collimation of the wind is due to density gradient effects in a disk around the star, the wind will be channeled into jets far away from the star, in which case the geometry sketched in Figure 4 will not apply, even in a rough approximation, to the H_α emission region. But the origin of wind collimation is still an issue, as is the collimation scale length (e.g. Torbett 1984).

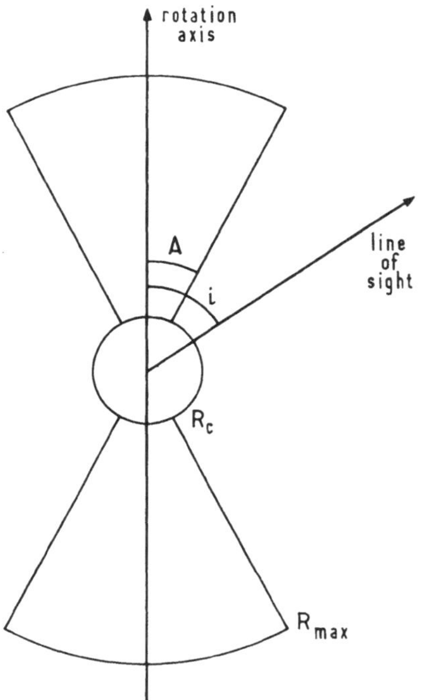

Fig.4 : Biconical envelope geometry used for computing the profiles of figs.5 to 7

Main envelope parameters are the cone's opening angle A and the envelope's maximum radius Rmax. The observer sees the envelope from an angle i. For exploratory purposes, I use a simple two-level atom model to compute the source function in the Sobolev approximation, and then integrate the profile exactly, using a generalization of the method described by Bertout (1984).

So far only radial, outward accelerating outflows have been investigated. This allows us to compare results with spherically symmetric computations and to study the role of the geometry alone. Parameters entering the computation of the two-level atom source function are

i) the radial optical depth, described by the customary parameter $\tau_o = (hc/4\pi) B_{12} N_c R_c / V_c$ where N_c is the number of lower state atoms and V_c the velocity at the stellar radius R_c.

ii) the stellar continuum intensity I_c and the Planck function B(r) at the local electron temperature. We assume $I_c = B = 1$.

iii) the ratio Σ of collisional to total de-excitation rates.

The velocity field is parametrized as

(1) $v(r) = V_c(r / R^c)^{\alpha}$

The number n_l of atoms in the lower state is further parametrized as

(2) $n_l = N_c(r / R_c)^{\beta}$

If conservation of the number of lower state atoms holds in the envelope, an assumption often made in parametrized computations, then $\beta = -(2 + \alpha)$. We note however that there is no physical reason for this assumption; while the total number of atoms will obey mass conservation in a steady-state flow, the radiative excitation conditions may lead to an utterly different distribution of the atoms in the line's lower state.

Figures 5 to 7 show the computed flux (in units of the underlying stellar continuum flux) as a function of the normalized frequency displacement $x = (\nu-\nu_o)/\Delta\nu_D$ for different values of α and β. In all cases, the cone's opening angle is A = 30°. In Figure 5, we have $\alpha = 0$ and $\beta = -2$, corresponding to a constant velocity outflow with conservation of the number of lower level atoms. Profiles corresponding to view angles of 30°, 60° and 90° are shown for $\tau_o = 10^2$ and $\Sigma = 10^{-2}$. The profile shape varies from a well-separated double peak structure at 30°, for which angle no gas with small radial velocity is seen by the observer, to a single, almost symmetrical peak centered at the line's nominal wavelength at 90°. The results in this simple case thus correspond well to what can be expected intuitively.

This simple picture changes, however, when one considers a radial velocity gradient, e.g. $\alpha = 1$. Figure 6 corresponds to the conservation of atoms in the lower state, i.e. $\beta = -3$, with $\tau_o = 10^2$ and $\Sigma = 10^{-2}$. Because of the strong density gradient outwards, only the inner parts of the flow contribute to the profile. At small view angles, this results in a classical P Cygni profile, while a moderate, two-peaked emission is seen at larger view angles. Relaxing the assumption of continuity for the lower state brings more expected results, as Figure 7 illustrates. There, we have $\beta = -1$, $\tau_o = 10^2$ and $\Sigma = 10^{-2}$. Now the complete volume of the cones contributes to the profile, and we get a double peaked structure with slightly blue-displaced minimum at low sight angles. As expected, the flux at central line frequencies increases as the sight angle increases and completely dominates the profile above about 70°.

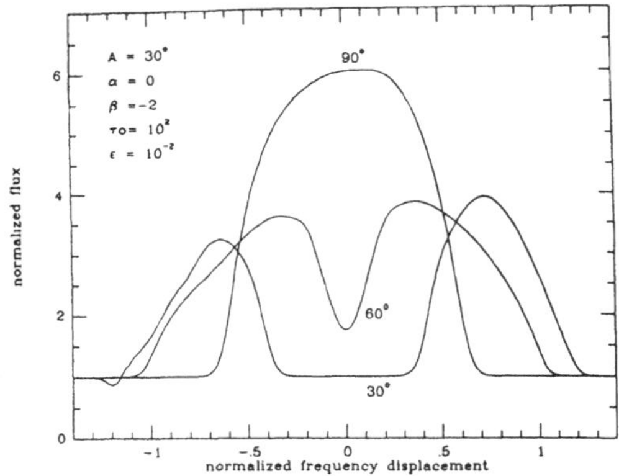

Fig. 5 : Computed profiles in the envelope geometry of Fig.4. Parameters of the computations (given in the left upper corner) are discussed in the text, and the profiles are labelled with the values of the view angle.

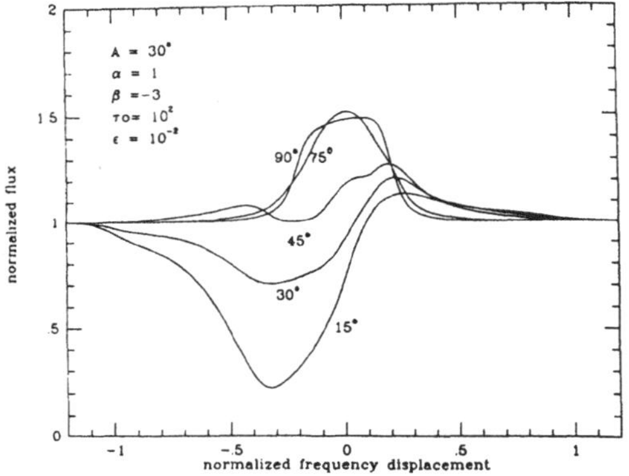

Fig. 6 : Same as Fig.5 but for a different velocity field (see text for details).

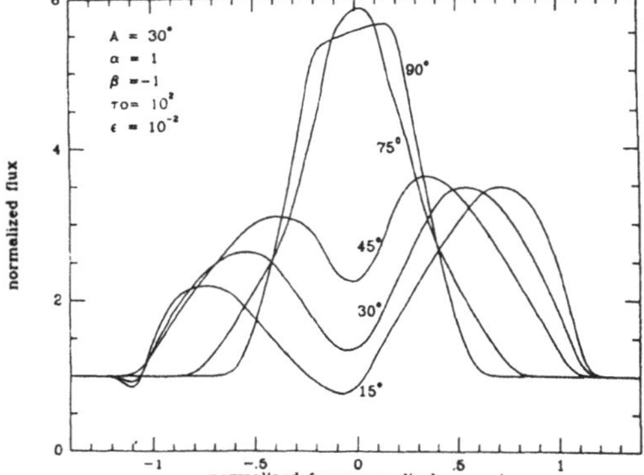

Fig. 7 : Same as Fig.6 but for a different density distribution (see text for details).

That the profile changes according to the slope of the density distribution appears to be a general feature of accelerated outflows. Although it makes it rather difficult to distinguish profiles formed in a biconical geometry from those formed in spherical geometry when the density of atoms in the lower level decreases faster than about r^{-2}, it may help when trying to derive a density distribution from profile shapes, assuming that the velocity field and flow geometry are approximately known by other means.

Despite the impossibility of offering any precise statements as a conclusion to this paper, for both the simple geometric model investigated here and the atom model are still too crude, the observed Hα profiles of Figures 1 to 3 do however display encouragingly striking qualitative similarities to the computed profiles of Figures 5 to 7. In particular, typical Type III P Cygni profiles are easily formed in biconical geometry. Because this property of the computed profiles is due to the bipolar nature of the flow rather than to details of the geometry, we feel confident that further studies of optical and molecular lines in various bipolar geometries based on sophisticated radiative transfer techniques will bring new insight to our understanding of mass-loss from T Tauri stars.

References

Ambartsumian V A 1947 Stellar Evolution and Astrophysics (Erevan, Acad. Sci. Armenian SSR)

Appenzeller I 1977 Proc. IAU Coll. No. **42** 80

Appenzeller I, Chavarria C, Krautter J, Mundt R and Wolf B 1980 Astron. Astrophys. **90** 184

Appenzeller I, Jankovics I and Oestreicher R 1983 submitted to Astron. Astrophys.

Appenzeller I, Jankovics I and Krautter J 1983 Astron. Astrophys. Suppl. **53** 291

Bally J and Lada C J 1983 Astrophys. J. **265** 824

Basri G, Bertout C and Calvet N 1984 in preparation

Bastien P 1981 Astron. Astrophys. **94** 294

Bastien P 1982a Astron. Astrophys. Suppl. **48** 153

Bastien P 1982b Astron. Astrophys. Suppl. **48** 513

Beals C S 1950 Publ. Dominion Astrophys. Obs. 9 1

Bertout C 1984 Rep. Prog. Phys. **47** 111

Bertout C, Carrasco L, Mundt R and Wolf B 1982 Astron. Astrophys. Suppl. **47** 419

Bertout C and Thum C 1982 Astron. Astrophys. **107** 368

Bieging J H, Cohen M and Schwartz P R 1984 Astrophys. J. **282** 699

Calvet N, Basri G, Imhoff C L and Giampapa M S 1984 submitted to Astrophys. J.

Calvet N, Basri G and Kuhi L V 1984 Astrophys. J. **277** 725

Calvet N, Canto J and Rodriguez L F 1983 Astrophys. J. **268** 739

Cohen M 1984 Phys. Reports in press

Cohen M, Bieging J H and Schwartz P R 1982 Astrophys. J. **253** 707

Cohen M and Kuhi L V 1979 Astrophys. J. Suppl. **41** 743

Cram L E, Giampapa M S and Imhoff C L 1980 Astrophys. J. **238** 905

DeCampli W M 1981 Astrophys. J. **244** 124

De Vegt C 1982 Astron. Astrophys. **109** L15

Edwards S 1985 in preparation

Edwards S and Snell R L 1982 Astrophys. J. **261** 151

Holzer T E, Fla T and Leer E 1983 Astrophys. J. **275** 806

Gahm G F 1981 Astrophys. J. Lett. **242** L163

Hanson R B, Jones B F and Lin D C 1983 Astrophys. J. Lett. **270** L27

Hartmann L, Edwards S and Avrett A 1982 Astrophys. J. **261** 279

Harvey P M, Thronson H A and Gatley I 1979 Astrophys. J. **231** 115

Herbig G H 1962 Adv. Astron. Astrophys. **1** 47

Herbig G H 1977 Astrophys. J. **214** 747

Herbig G H and Rao N K 1972 Astrophys. J. **174** 401

Jones B F and Herbig G H 1979 Astron. J. **84** 1872

Jordan C, de Ferraz M C and Brown A 1982 Proc. 3rd Europ. IUE Conf. ESA Publ ESA-SP 176, p.83

Joy A H 1945 Astrophys. J. **102** 168

Krautter J and Bastian U 1980 Astron. Astrophys. **88** L6

Ku W and Chanan G A 1979 Astrophys. J. Lett. **234** L59

Kuhi L V 1964 Astrophys. J. **140** 1409

Kuhi L V 1978 Protostars and Planets ed. T Gehrels (Tucson, University of Arizona Press) p. 708

Lago M T V T 1979 DPhil Thesis University of Sussex

Lago M T V T 1984 preprint

Mundt R 1984 Astrophys. J. **280** 749

Mundt R and Fried J W 1983 Astrophys. J. Lett. **274** L83

Mundt R and Hartmann L 1983 Astrophys. J. **268** 766

Norman C and Silk J 1980 Astrophys. J. **238** 158

Rydgren A E, Strom S E and Strom K M 1976 Astrophys. J. Suppl. **30** 307

Schmid-Burgk J 1982 Astron. Astrophys. **108** 169

Silk J and Norman C 1983 Astrophys. J. Lett. **272** L49
Simon T, Schwartz P R, Dyck H M and Zuckerman B 1983 IAU Coll. No.71 p.505
Torbett M V 1984 Astrophys. J. **278** 318
Ulrich R K and Knapp G R 1979 in preparation
Vogel S and Kuhi L V 1981 Astrophys. J. **245** 960
Wagenblast R, Bertout C and Bastian U 1983 Astron. Astrophys. **120** 6
Walker M F 1956 Astrophys. J. Suppl. **2** 365
Walker M F 1972 Astrophys. J. **175** 89
Walter F M and Kuhi L V 1981 Astrophys. J. **250** 254
Zappala R R 1972 Astrophys. J. **172** 57

MASS OUTFLOWS FROM T TAURI STARS AND THEIR INTERACTION WITH THE ENVIRONMENT

Reinhard Mundt

Max-Planck-Institut für Astronomie
Königstuhl, D-6900 Heidelberg, Fed. Rep. of Germany

ABSTRACT

A review of the various observational indications for mass outflows from visible T Tauri stars and IR-sources of similar luminosity ($0.1\ L_\odot \lesssim L \lesssim 100\ L_\odot$) is given. Stellar emission line profiles, radio continuum studies, CO line observations of surrounding high-velocity molecular gas, and optical investigations of associated Herbig-Haro (HH) objects and optical jets are discussed. The latter investigations have shown that some of these young stellar objects are capable of generating highly-collimated bipolar flows. These jets have typical lengths of 0.01 to 0.2 pc, opening angles of about 3 to 10 deg, and their measured radial velocities range from 50 to 400 km/s. The properties of these jets and their relation to HH objects are detailly discussed. A model is proposed in which HH objects are powered by jets emanating from nearby young stars.

I. INTRODUCTION

A large and growing body of observational material indicates that energetic and bipolar mass outflows are an important phase in early stellar evolution for probably all types of young stars. This refers to low-mass stars ($m \approx 1\ m_\odot$) of a few solar luminosities, like the T Tauri stars, as well as to more massive objects having 10^5-times higher luminosities. In the first part of this review we discuss the various observational evidence for mass outflows from (visible) T Tauri stars and IR-sources of comparable luminosities. Within the recent years many observational data have been collected in this field. Many of them through CO-line observations at mm wavelengths, but also through observations at optical, infrared and cm wavelengths. For example, optical line profile studies of the emission lines of many T Tauri stars and related objects have significantly improved our knowledge of the wind properties near the stellar surface. Furthermore, observations at mm wavelengths of the molecular gas near these stars and optical studies of associated HH objects have been very important in studying the geometry of their outflows at larger distances ($\approx$ 0.01-1 pc). Those observations showed that these flows are often bipolar and well collimated. In many cases the high flow collimation is directly evident from the appearance of optical high-velocity (100-400 km/s) jets extending from these stars to distances of up to 0.2 pc. In the second part of this review we will discuss in detail the properties of these highly-collimated flows. This section will be less comprehensive than a recent review by the author on this topic (Mundt 1984a).

For a more general discussion of HH objects the reader is refered to the reviews of Böhm (1983), Canto (this volume), and Schwartz (1983). A recent review on T Tauri stars has been given by Bertout (1984) and various aspects of their winds have been discussed by Hartmann (1984b). The properties of the CO flows near HH objects and their exciting stars has recently been described by Edwards and Snell (1984, and references therein). An important reference for all these topics are the proceedings of the Symposium on HH objects and T Tauri stars held in Mexico City in February 1983 to honor Dr. G. Haro (Vol. 7, 1983 of Rev. Mex. Astr. Ap.)

II. OBSERVATIONAL EVIDENCE FOR MASS OUTFLOWS

II.1 Optical Line Profile Studies

It has been known for several decades that some T Tauri stars and related objects like the Herbig Ae/Be stars show evidence for outflowing matter in their line profiles (Joy 1945, Herbig 1960, Kuhi 1964,). Within recent years the line profile data of these objects have enormously increased, mainly through the availability of sensitive detectors (Herbig 1977, Hartmann 1982, Bertout et al. 1982, Mundt and Giampapa 1982, Ulrich and Knapp 1984, Mundt 1984b, Finkenzeller and Mundt 1984, Bastian and Mundt 1985). These observations have mainly been carried out at Hα and the NaD lines. The line profiles indicate that the mass motions in the envelopes of T Tauri stars are rather complex and for example, mass infall and outflow may occur simultaneously in one star (see e.g. Bertout 1984 or Mundt 1984b).

The number of T Tauri stars which show direct evidence for mass outflow in their emission lines seems to be relatively small. According to Kuhi (1978) only about 5% out of 70 stars show classical P Cygni profiles in Hα , while most stars (≈ 60%) show a blueshifted "absorption reversal" not reaching below the adjacent continuum. This "absorption reversal" may be caused by a stellar wind as well (see Bertout 1984 for a detailed discussion). However, Mundt (1984b) has shown that the fraction of stars showing direct evidence for outflowing matter is much higher among the strong-emission T Tauri stars, of which many show no detectable photospheric absorption spectrum (often classified as "continuous"). In a sample of 11 stars with $W_\lambda(H\alpha) \gtrsim 80$ Å he found evidence for mass loss in Hα or (and) NaD in 9 stars. In the Cohen and Kuhi (1979) catalogue about 30% of the T Tauri stars in the Taurus-Auriga region have such strong Hα lines. This means, the fraction of T Tauri stars in this region, for which direct evidence for mass loss is expected to be found by high-quality spectra is in the order of 20-30%. According to Cohen and Kuhi (1979) the T Tauri stars in Taurus-Auriga belong to the youngest stars and have the highest portion of strong-emission stars among all T-associations they investigated. This suggests that on the average the strong-emission stars are younger. If the strength of the emission lines in T Tauri stars is a mass loss rate indicator, as argued by Mundt (1984b), their mass loss is apparently decreasing already in a fraction of a typical T Tauri age (10^5-10^6 yr).

Fig. 1 shows examples of P Cygni-profiles observed in the Hα and NaD lines of the strong-emission stars AS 353A und DG Tau, respectively. Both stars are the "exciting" stars of HH objects and optical jets. Interestingly, similar indications for mass loss have been found in all other stars of this subgroup, for which high-quality spectra have been obtained (Mundt, Brugel, and Bührke 1984).

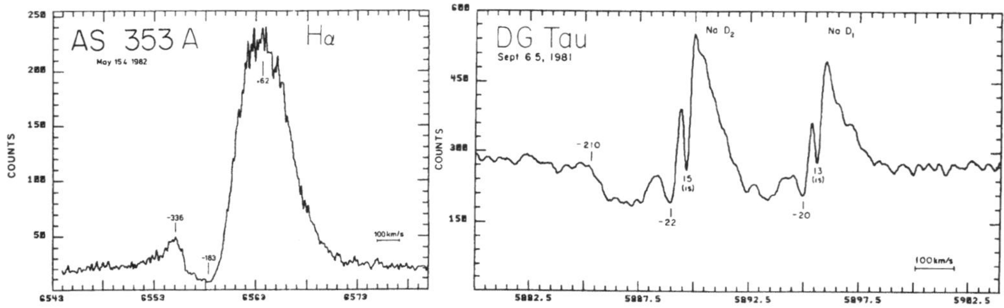

<u>Fig. 1:</u> P Cygni profiles observed in the Hα and NaD lines of AS 353A and DG Tau, respectively (figure from Mundt 1984b).

Important stellar wind parameters are the terminal velocity v_∞ and the mass loss rate $\dot{M}$. Both parameters, in particular the mass loss rate, are poorly known. The $\dot{M}$ values derived from optical line profiles range from 10^{-9} - 10^{-7} $M_\odot/yr$ (with considerable uncertainties, e.g. Bertout 1984). One can estimate v_∞ by measuring the maximum velocities occuring in the forbidden lines of these stars or in their associated HH objects (see also Appenzeller et al. 1984). The derived values of v_∞ range from 100-400 km/s. However, in some stars the situation seems to be less clear and their terminal velocities may be only 50-100 km/s (Mundt 1984b). Finally, it should be pointed out that wind properties similar to those of some T Tauri stars are observed in quite different young stellar objects as well. Examples are the FU Orionis stars and some Herbig Ae/Be stars, like Z CMa. In these objects a common wind acceleration mechanism seems to operate which produces cool ($T \lesssim 5000$ K) winds having velocities of 200-400 km/s already very close to the stellar surface ($R \lesssim 1.5R_*$, Bastian and Mundt 1985, Mundt 1984b).

II.2 Radio Continuum Observations

About 15 T Tauri stars and IR sources of comparable luminosities have been detected within the recent years at cm wavelenths mainly by utilizing the VLA (Bieging, Cohen, and Schwartz 1984, Brown and Mundt 1984, Bertout 1984 and references therein). Many investigators have interpreted their radio data by free-free emission from a stellar wind. By assuming terminal velocities of several hundred km/s, $\dot{M}$ values of 10^{-8} - 10^{-7} $M_\odot/yr$ have been derived.

This interpretation has been criticised by Mundt and Hartmann (1983). They pointed out that the radio emission may be of nonthermal origin. For example, caused by flares as observed in other late -type active stars (e.g. RS CVn variables, Gibson 1984). They showed that the large radio fluxes of some stars (e.g. V410 Tau) are difficult to reconcile with their optical properties. In the case of V410 Tau recent observation indicate, that it is a radio variable (Bertout 1984, Bieging, Cohen, and Schwartz 1984), giving further support for the idea that its radio emission is produced during flare-like events. An additional example of a radio-variable T Tauri-like star has been found by Feigelson and Montmerle (1984, see also T. Montmerle this volume). Furthermore, there are indications for flare-like activities in T Tauri stars at X-ray wavelengths (e.g. Montmerle et al. 1983). It would not be surprising if the same type of activity is responsible for the radio emission in some of these stars, since correlations between the X-ray and radio emission are suggested for some late-type active stars (Drake, Simon, and Linsky 1984). A further difficulty in the interpretation of the radio data are the spectral indices α, which are consistent only in a few cases with a stellar wind having an electron density distribution $\sim 1/r^2$ (α = 0.6). In some cases may be even negative, which would imply a non-thermal origin of the stellar wind as well.

In summary, it appears that at present there are good reasons to question the simple picture of free-free emission originating in a dense wind as the explanation of radio emission from all T Tauri stars. For those stars in which the radio emission is produced that way only lower limits on the mass loss rate can be derived, as long as the degree of ionization in the wind is unknown. As indicated by their NaD line profiles, many T Tauri stars have wind regions of rather low temperature ($T \lesssim 5000$ K). However, it will be shown in Sec. III.3 that high resolution radio continuum observations can provide important information on the geometrical distribution of the ionized gas near these stars.

II.3 CO-Line Observations

Within recent years the mass outflows from young stars have been heavily investigated by observations of high-velocity molecular cloud components probably being accelerated by the winds from these objects. These investigations envolved the measurement of CO emission line profiles in most cases. In these profiles the high-velocity molecular gas (HVMG) accelerated by these stars is indicated by broad line wings. For T Tauri stars and IR-sources of comparable luminosity the CO wings (if detected) extend to velocities of 5-30 km/s. In several cases bipolar mass motions are indicated (e.g. Snell, Loren and Plambeck 1980, see also Fig. 2 and 6). So far, searches for HVMG have been carried out for about 50 visible T Tauri stars and related objects (Edwards and Snell 1982, Kutner et al. 1982, Calvet, Canto, and Rodriguez 1983). Furthermore, such searches have been performed in the vicinity of about 40 HH objects and their exciting stars (Edwards and Snell 1984, and references therein). These stars are in most cases IR sources deeply embedded in molecular material having similar luminosities as visible T Tauri stars.

For reasons of homogeneity we will discuss the data given by Edwards and Snell (1982, 1984) only. Among visible T Tauri stars they

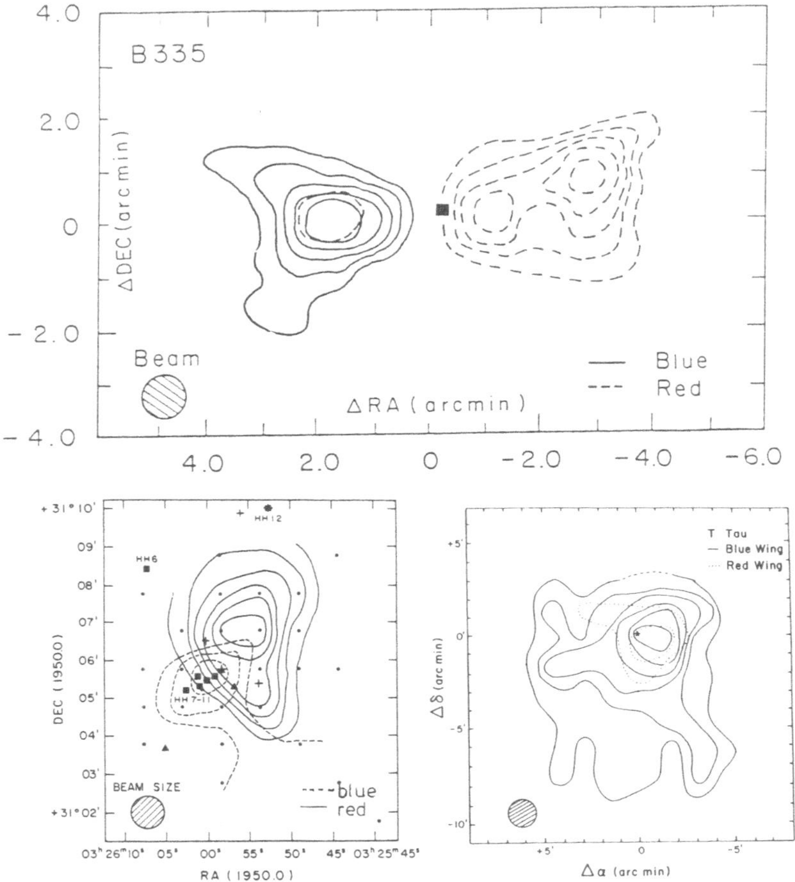

<u>Fig. 2:</u> Distribution of the redshifted and blueshifted high-velocity (5-30 km/s) CO-gas near 3 young stellar objects of low to medium luminosity (10-100 $L_{\odot}$). B335 is an example of distinctly bipolar molecular flow, while for T Tau there is no hint for such a flow pattern at all and predominantly blueshifted CO gas is observed. The CO flow associated with HH 7-11 can be regarded as an intermediate case (figures from Goldsmith et al. 1984, Snell and Edwards 1981, and Knapp and Padgett 1984, respectively).

found only three cases (out of 25) with associated HVMG. These cases are T Tau, AS 353A and B, and XZ/HL Tau. T Tau (or its IR-Companion), AS 353A, and HL Tau are known to be the exciting stars of HH objects or optical jets (Bührke, Brugel, and Mundt 1985, and Sec. III, below). However, about 75% of the surveyed HH objects have been found to be in the vicinity of a molecular flow. This means that the exciting star of an HH object is often also the driving source of a molecular flow.

The stars with associated HVMG and HH objects are believed to belong to the youngest pre-main-sequence objects, since in most cases they are not visible stars but IR objects still deeply embedded in (circumstellar) gas and dust. Why do we observe in the vicinity of these objects phenomena like HH objects or bipolar CO flows and not for visible T Tauri stars? One reason might be higher mass loss rates in the earliest phases of their stellar evolution, as alreadys discussed above. On the other hand it has been argued by Edwards and Snell (1982) that at least some visible T Tauri stars have apparently similar mass loss rates like the deeply embedded sources, but are probably surrounded by such a tenous molecular gas environment that any stellar wind interactions with it are not detectable. It will be difficult to decide which of these two scenarios might be correct, as long as there are no independent and reliable means to measure mass loss rates for visible as well as for highly obscured young stars and as long as it has not been proved that high-velocity stellar winds are actually driving the CO-flows (see also below).

In some cases however, it is indeed indicated that the molecular cloud environment strongly influences the CO flow pattern and it is therefore conceivable that similar effects cause the absence of detectable HVMG near some objects. An example is T Tau (see Fig.2). In this case one observes predominantly blueshifted CO gas. This is probably caused by strong density gradients in the environment of T Tau with the more tenous gas (which is easier to accelerate) located towards the earth (Knapp and Padgett 1984).

CO maps of the HVMG are available so far for about 15 T Tauri stars and IR-sources of comparable luminosity (Edwards and Snell 1984 and references therein,Leverault 1984, Goldsmith et al. 1984). About half of these flows have a bipolar structure. The flow from L1551-IRS5 is one of the best examples (Fig. 6, below). A more recent example (B335) is shown in Fig. 2. The bipolar CO-flow pattern of L1551-IRS5 and other sources have been interpreted by a collimated high-velocity ($\approx$ 200 km/s) bipolar wind accelerating the surrounding molecular gas (Snell, Loren and Plambeck, 1980). As discussed in Sec. III there is indeed various observational evidence that such jets are emanating from at least some of these objects. However, there are severe energetic problems in accelerating the CO gas by high-velocity stellar matter. If the HVMG is accelerated by a wind with $v_W \approx$100-400 km/s then mass loss rates can be derived from the CO data by assuming momentum conservation. For sources of $L \approx$ 1-100 $L_\odot$ the derived $\dot{M}$ values are very large and range from 10^{-8} $M_\odot$/yr to several 10^{-6} $M_\odot$/yr. The wind luminosities $L_W = 0.5\ \dot{M}V_W^2$ are in several cases on the order of the stellar luminosity ($L_W/L_* \approx 0.01$-0.5). This poses severe energy problems for all wind acceleration mechanisms proposed for these objects (e.g. Hartmann 1984a). Due to these problems in generating energetic high-velocity winds it has been suggested that the CO outflows are not driven by such winds but by the gravitational and magnetic energy of matter accreting onto these objects (Draine 1983, Pudritz and Norman 1983, Uschida and Shibata 1984).

II.4 HH Objects and Optical Jets

It is now generally accepted that the emission lines of HH objects originate in the cooling regions of high-velocity ($v \approx 100$ km/s) shockwaves (e.g. Böhm 1983), which probably arise from the dynamical interaction of the flows from young stars with the ambient medium. HH emission is therefore well suited to probe the regions close to these stars for the distribution of the high-velocity matter directly ejected by them. Recent proper motion and radial velocity measurements of HH objects suggest that this matter is often channeled into well-collimated bipolar flows. In many cases the high flow collimation is directly indicated by the appearance of narrow high-velocity jets, extending from T Tauri stars and related young stellar objects. As discussed in Sec. III.4 below, these jets and the HH objects are related phenomena, which is evident from their similar spectra and from the physical association observed in several cases.

In the following section we discuss in more detail the properties of highly-collimated (bipolar) flows from T Tauri stars and IR-sources of comparable luminosity. This will be done mainly on the basis of optical studies of HH objects and jets. As mentioned above, a more comprehensive review on this topic has recently been given by Mundt (1984a) and therefore only some aspects will be discussed here.

III. OPTICAL STUDIES OF HIGHLY COLLIMATED BIPOLAR FLOWS

III.1 Observational Evidence

The first example of a bipolar flow from a young star being traced by HH emission only was discovered in the case of HH 1 and HH 2 by Herbig and Jones (1981). Through proper motion measurements they showed that the individual knots in HH 1 and HH 2 are moving in opposite directions with tangential velocities of 100-300 km/s, suggesting a highly anisotropic bipolar flow being driven by a star located in between HH 1 and HH 2. A few years before these proper motion have been measured, Cohen and Schwartz (1979) had found a T Tauri star near the connecting line of HH 1 and 2, which might be the driving source of this flow (see also Mundt and Hartmann 1983).

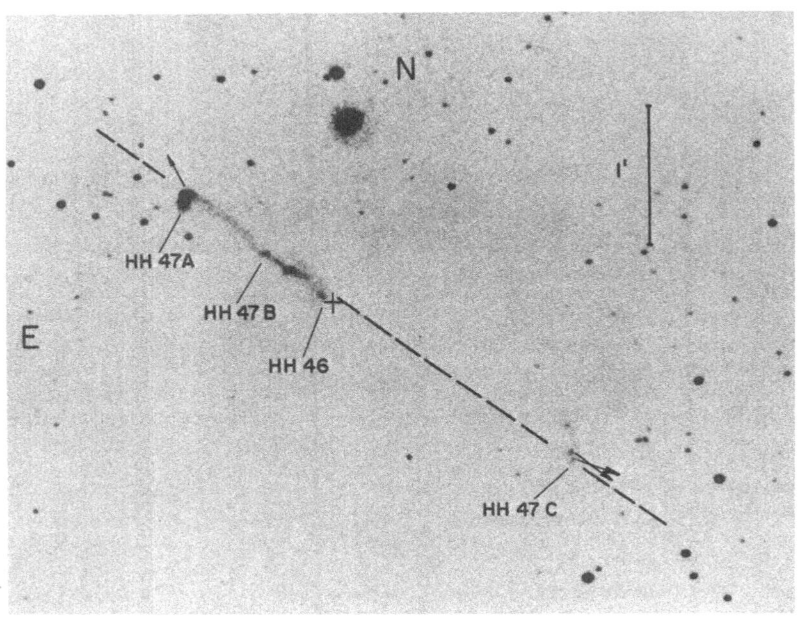

Fig. 3: [SII]-6716,6731-line image of the HH 46/47 region. The IR-source driving this bipolar flow is indicated by a cross (figure from Graham and Elias 1983). The arrows show the proper motion of HH 47A and HH 47C within 500 yr (from Schwartz, Jones and Sirk 1984).

One of the most remarkable cases for a well-collimated bipolar flow is HH 46/47 (Dopita, Schwartz, and Evans 1982; Graham and Elias 1983). An image of the HH 46/47 region in the [SII] lines is reproduced in Fig. 3. One observes a 1 arcmin long jet with HH 46, HH 47B and in particular HH 47A being its brightest portions. Similar examples will be given below. In this case all the material in the jet is blueshifted ($v \approx -100$ to -190 km/s), while two small HH knots (HH 47C) located 1.5 arcmin southwest of HH 46 are redshifted ($v \approx +100$ km/s). The low luminosity IR-source ($L \approx 12L_\odot$) driving this bipolar flow is located near the southwest end of the jet (see Fig. 3). Radial velocity masurements indicate higher velocities in the jet than in the bright HH knot HH 47A, suggesting a deceleration of the flow in this knot, which apparently represents its termination point.

Bipolar mass motions similar to those observed for HH 1/2 and for HH 46/47 were discovered by Mundt, Stocke and Stockman (1983) for the HH objects HH 32A-D associated with the T Tauri star AS 353A. A deep Hα CCD frame, recently obtained of these objects, shows that HH 32A, B and D are the brightest knots of a jet extending from AS 353A towards HH 32A where it apparently terminates (akin to the HH 46/47 jet). All the material in the jet is highly redshifted ($V_{max} \approx 400$ km/s) while the compact knot HH 32C, on the opposite side of AS 353A, is highly blueshifted (see Mundt 1984a, and Stocke, Hartigan, and Mundt 1984 for a detailed discussion).

Jets similar to those observed in the case of HH 46/47 and AS 353A were recently discovered by Mundt and Fried (1983) and Mundt et al. (1984) by deep CCD imaging of T Tauri stars and HH objects utilizing the 2.2 m telescope on Calar Alto, Spain. Altogether, they obtained deep CCD images of about 35 strong-emission T Tauri stars and about 15 HH objects. In order to discriminate HH nebulosities from continuum reflection nebulosities they used filters which either <u>included</u> or <u>excluded</u> the strongest HH emission lines (like Hα, [NII] , [SII]). In total they discovered 7 jet-like HH (or Hα) nebulosities. These jets either extend from T Tauri stars and low-luminosity IR-sources or they apparently point towards bright HH objects. Several of these newly discovered jets are displayed in Fig. 4. Since the discovery of these jets radial velocities have been measured for most of them (Mundt, Brugel, and Bührke, 1984). The values obtained range from 50-250 km/s (see Table 1).

As already discussed by Mundt and Fried (1983) and Mundt et al. (1984) it is strongly indicated that these jets are all representations of well-collimated high-velocity flows emanating from the immediate vicinity of their associated stars. This interpretation is strongly supported not only by the morphology of the jets, but also by their high velocities of several hundred km/s. In addition, these high velocities are also indicated by their HH-type spectra (observed in all cases which were spectroscopically studied so far). This requires shock heating of the emitting material with velocities of about 100 km/s. Furthermore, it follows from their HH spectra that these jets can be regarded as highly elongated HH objects, but with much lower surface brightnesses than most "classical" HH objects (see also Sec.III. 4).

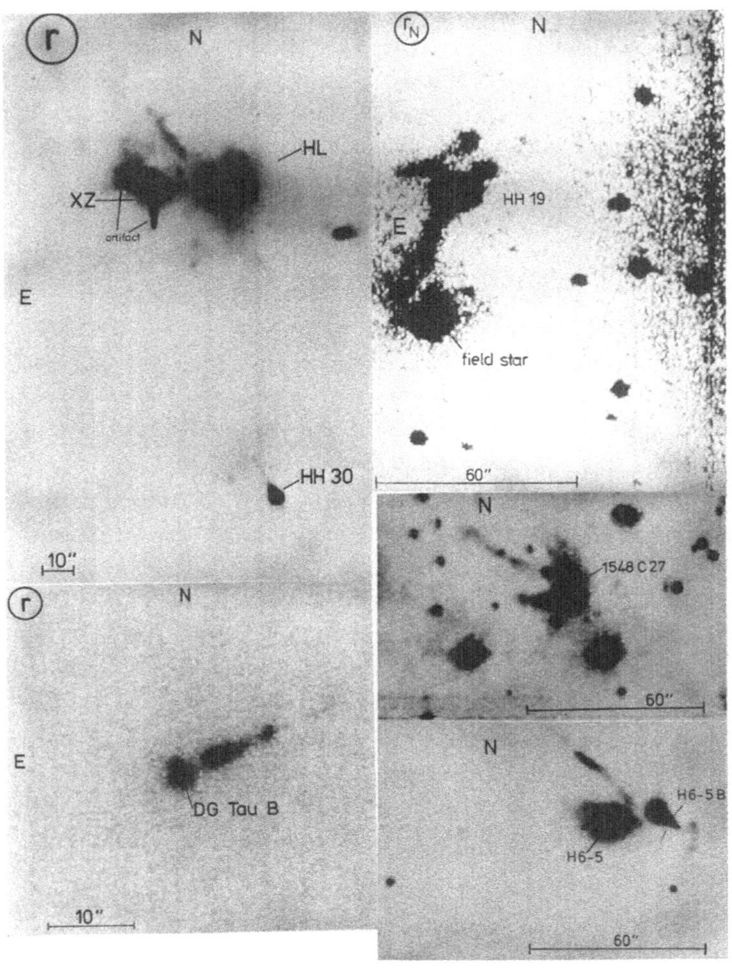

Fig. 4: Examples of jets extending from T Tauri stars and low-luminosity IR-sources or being associated with bright (previously known) HH objects (from Mundt and Fried 1983, and Mundt et al. 1984).

The fraction of visible T Tauri stars being associated with HH objects or optical jets can be estimated to be in the order of 5%. With the exception of the Cohen-Schwartz star all of these stars have strong emission lines and are therefore apparently younger (see above, and also Mundt, Brugel and Bührke 1984). In addition, most of the "exciting" stars of HH objects and molecular flows are low-luminosity IR-sources, which appear to be younger than visible T Tauri stars. In summary, it seems that the outflow activities detectable in the form of CO-flows, HH objects, and optical jets occur only in the very earliest phases of evolution of young low-mass stars (e.g. at a age of 10^4 - 10^5 yrs).

Table 1: Properties of Highly Collimated Flows as Derived From Optical Studies of Jets and HH Objects

Associated stars or HH Objects	Distance (pc)	Length (arc sec)	Length (pc)	Approximate Opening Angle (degrees)	Radial Velocity (km/s)	Proper motion[b] (km/s)	Luminosity of Driving Source ($L_\odot$)	Ref.
HH 12	350	120	0.2	5-10	-100	60-250 (12)	-	1,2
HH 7-11	350	65	0.1	10	-30 to -150	60 (11)	58	2,3,4
HL Tau	150	20	0.02	5	-170	-	7.2	4,5,6
HH 30	150	13	0.01	5	-	-	-	5
L1551-IRS 5	150	17	0.015	10	-80 to -190	150(28), 170(29)	32	4,5,6,7,8
DG Tau	150	8	0.006	10	-150 to -250	-	7.6	5,6
DG Tau B	150	17	0.015	5	-	-	-	5
Haro 6-5 B	150	60	0.05	5-10	-40[a] +50[a]	-	-	6,9
HH 1/2	460	55/110	0.12/0.25	10	-8[a] +17[a]	155-350 (1 A-F) 60-290 (2 A-I)	27	4,10
HH 33/40	460	75	0.17	5	+80 to +160	-	-	6,9
HH 19	460	30	0.07	-	+20	-	-	6,9
R Mon/HH 39	800	15/400	0.06/1.6	10	-75[a] +170[a]	60 (39 D) 275-310 (39 A,C,E)	1360	4,11,12
HH 46/47	400	75	0.15	5	-110 to -190[a] +120[a]	160 (47 A) 230 (47 C)	12	13,14,17
AS 353 A	300	29	0.035	10	-20 to -350[a] +60 to +380[a]	50 (32 A) 200 (32 B)	6.6	2,4,15,16
1548C27	?	45	$0.1 \frac{d}{500\ pc}$	3	-175	-	-	6,9

a = optically bipolar flow, b = number in parenthesis gives HH Object number

1= Strom, Strom, and Stocke 1983, 2=Herbig and Jones 1983, 3=Snell and Edwards 1981, 4=Cohen et al. 1984, 5= Mundt and Fried 1983, 6=Mundt, Brugel, and Bührke 1984, 7=Sarcander, Neckel, and Elsässer 1984, 8=Cudworth and Herbig 1979, 9= Mundt et al. 1984, 10=Herbig and Jones 1981, 11=Jones and Herbig 1982, 12=Brugel, Mundt,and Bührke 1984, 13=Graham and Elias 1983, 14=Emerson et al. 1984, 15=Mundt, Stocke, and Stockman 1983, 16=Stocke, Hartigan, and Mundt 1984, Schwartz, Jones, and Sirk 1985

III.2 Lenght Scales, Velocities and Opening Angles of Highly-Collimated Flows

Table 1 summarizes the properties of all presently known highly-collimated flows (in total 14) as derived from optical studies of jets and HH Objects. The typical lenghts of these flows range from 0.01 pc to 0.2 pc (or 10"-100" on the sky) and the opening angles have values between ~ 3° and ~ 10°. This means that these flows are much more collimated (at least close to their driving stars) than the molecular flows observed in CO. Unfortunatly, most of the CO observations are of too low spatial resolution to allow a reasonable comparison with the optical data. The measured radial velocities and proper motions given in Table 1 indicate flow velocities of 100-400 km/s (without correcting for projection effects or for the deceleration in shock fronts). In several cases (e.g. HL Tau, DG Tau B, HH 30) the jet is spatially unresolved perpendicular to its axis ($\lesssim$ 2"), providing upper limits on its diameter of $\approx 4\text{x}10^{15}$ cm (for r = 150 pc). Nearly all driving stars of these flows are of low to moderate luminosity ($L \approx$ 1-100 $L_{\odot}$). An exception is R Mon (Brugel, Mundt and Bührke 1984), which has a luminosity of $\approx$ 1400 $L_{\odot}$. It is the only luminous object for which optical studies found evidence for a well-collimated bipolar flow (see Fig. 5). We note however that very little effort has been made to find jets around high-luminous objects similar to those described above. Such detections would be very important. However, these objects are more difficult to find, since on the average they are more distant, higher reddened, and often associated with bright reflection nebula.

III.3 Collimation Length Scales and Collimation Mechanisms

The emission line CCD frames of L1551-IRS5, DG Tau B and HH 30 show that these flows are already collimated on the resolution limit of these frames of 1-2", which means at length scales of a few 10^{15} cm (Mundt and Fried 1983). For L1551-IRS5 and DG Tau the VLA maps of Bieging, Cohen, and Schwartz (1984) shown in Fig. 6 and 7 indicate even smaller collimation length scales. The innermost radio contours (resolution 0.5") are already elongated in a similar direction as the optical jets of these two objects. The upper limit on the collimation length derived from these VLA data is about 10^{15} cm (or $\approx$ 70 AU). The small collimation length scale of DG Tau has recently been confirmed optically by long-slit spectroscopy (Mundt, Brugel, and Bührke 1984).

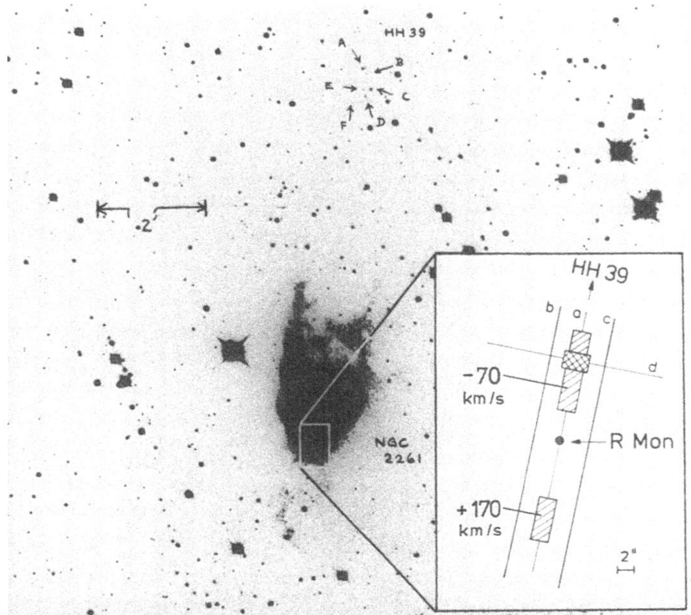

Fig. 5: R Mon/NGC 2261-region. The right inserted part of this figure shows where high-velocity [SII] 6716, 6731 - line emission has been detected near R Mon with the help of 4 long-slit spectrograms placed at positions a-d. As indicated in this figure [SII] line emission occurs in two regions only, which are located on the opposite sides of R Mon and are moving into opposite directions along the line connecting R Mon and HH 39 (from Brugel, Mundt, and Bührke 1984).

These spectra show, that in Hα a high-velocity component of about V = -185 km/s extends in the direction of the jet which can be traced already within 1" from the star (see Fig. 7).

The small collimation length scales derived here for the high-velocity ($\approx$ 100 km/s) flow components do not support any model by which the molecular flows are focused by large-scale (10^{17}-10^{18} cm) density gradients (e.g. Torrelles et al. 1983). Although the molecular flows might not be accelerated by high-velocity stellar winds (Sec. II.3), it seems very unlikely that they are focussed on such large scales.

A number of schemes have been proposed to channel the outflows from young stars into two opposite directions (e.g. Königl 1982, Pudritz and Norman 1983, Ushida and Shibata 1984). In the Königl model the energy source for all outflow phenomena is an isotropic high-velocity stellar wind, while in the other two models it is

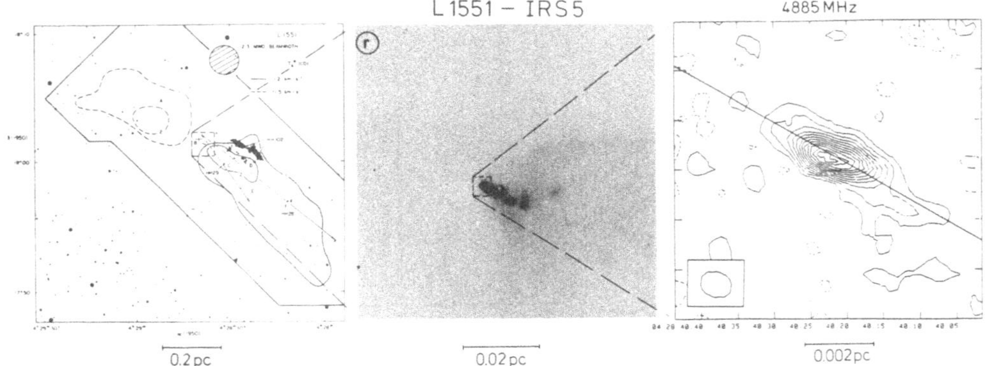

Fig. 6: L 1551-IRS5 flow. Left panel: Optical image of the L 1551 dark cloud with the bipolar high-velocity CO gas lobes superposed (figure from Snell, Loren, and Plambeck 1980). Middle panel: CCD frame of the IRS5 region showing the jet extending away from the position of the IR-source (figure from Mundt and Fried 1983). Right panel: 6 cm VLA radio continuum map of IRS5 (figure from Bieging, Cohen, and Schwartz 1984). Note the alignment of the flow in the same direction over a length scale range of 3 orders of magnitude.

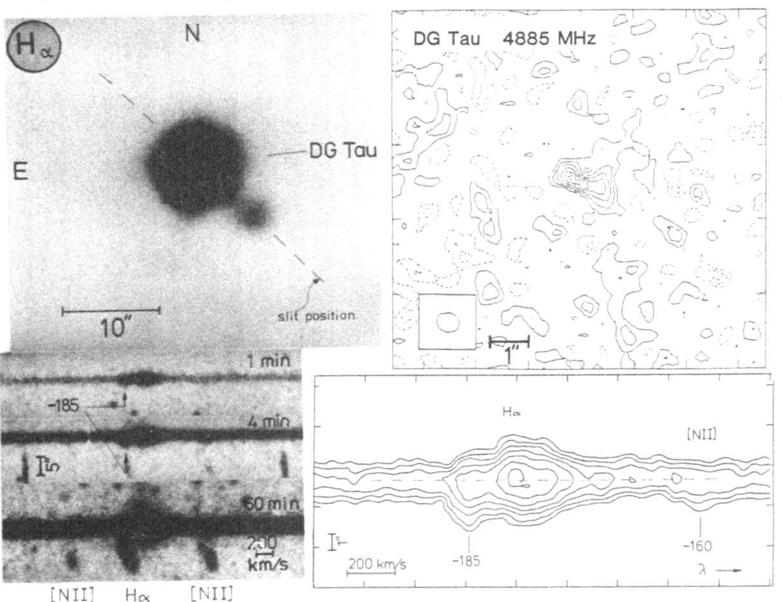

Fig. 7: Top left panel: Hα CCD frame of DG Tau showing a 8" long jet with an HH knot at the end (from Mundt and Fried 1983). Top right panel: 6 cm VLA radio continuum map of DG Tau. An extended, 1.5" long radio emission region is observed being oriented in nearly the same direction as the optical jet (from Bieging, Cohen, and Schwartz 1984). The lower panel shows differently exposed long-slit spectrograms with the slit oriented along the jet together with a denstiy contour plot of one spectrogram. These latter data show that in Hα a high velocity component (V = -185 km/s) is protruding from DG Tau into the direction of the jet, which is detectable already on scales of ≈1" (Mundt, Brugel, and Bührke 1984).

gravitational energy from an accretion disk. As discussed in Sec. II.1 optical observations, indeed suggest the presence of strong winds emanating from the surfaces of some T Tauri stars. The Königl model requires that these winds blow an energy driven bubble into the surrounding medium (disk). However, the calculations of Dyson (1984) show that such a bubble will within a few hundred years develop into a momentum driven one, due to strong radiation losses in the relatively dense bubble (for V_w = 300 km/s, $\dot{M} = 10^{-7}$ $M_\odot$/yr, and an ambient density of 10^4 cm^{-3}). To avoid these difficulties V_w-values of about 500 km/s are required, since the time scale t_e for changing an energy driven bubble into a momentum driven one is extremely sensitive to V_w ($t_e \sim V_w{}^{21}$, Dyson 1984). Such V_w-values would be much higher than the typical values expected for T Tauri stars of 100-300 km/s.

In all models proposed for the generation of bipolar flows a circumstellar disk is envolved. Is there any observational evidence for disks around the jet sources described above or for any other bipolar flow source? In a number of such sources (e.g. L1551-IRS5, DG Tau) one observes a high linear polarisation (6-20%) with the polarisation vector oriented (nearly) perpendicular to the flow direction (Mundt and Fried 1983, Hodapp 1984). As discussed by these authors these polarisation data can in principle be explained by scattering of stellar light in the polar lobes of an optically thick disk (Elsässer and Staude 1978). Other studies suggesting the presence of disks around these jet sources are the CS-line observations of L1551-IRS5 by Kaifa et al. (1984, but see also Wamsley this volume)

III.4 The Powering of HH Objects by High-Velocity Jets

In Table 1 about 10 cases are listed for which stellar-like HH objects form the brightest parts ("hot spots") of a jet. In some cases the jet apparently terminates at the position of an HH object (HH47A, HH33), while in the case of HH 7-11 a chain of roughly equally spaced HH knots is observed. The latter morphology is found for other jets as well (e.g. L1551-IRS 5).

As discussed above, the observed jets seem to be indeed representations of quasi-continuous, well-collimated high-velocity flows (see also Mundt 1984a). Therefore, an HH object model with the following basic concept is suggested, at least for those HH objects associated with optical jets: HH objects trace the locations of high energy dissipation in the (strongest) shock fronts of high-velocity

(200-400 km/s) jets. Again, this is only the basic idea of this model and it has to be pointed out that in the inhomogenous environment of these jets many different mechanisms can cause the occurence of strong shocks (knots) in these flows and furthermore very complex flow pattern probably develop. An example of a mechanism which might be important in this environment is the pressure reconfinement of a free jet by the ambient gas (e.g. in regions with small density gradients, see Sanders 1983). The various formation mechanisms for such strong shocks (knots) in astrophysical jets have been discussed comprehensively by Normann, Smarr, and Winkler (1984). They showed that in addition to various external influences these shocks can be excited by numerous fluid dynamical instabilities (e.g. pinching). According to their "numerical jet experiments" the shock systems caused by fluid instabilities are not stationary and the pattern velocity for the individual shocks (knots) can be quite different. These effects might explain why in the HH 7-11 flow only the HH 11 knot has a high proper motion (Herbig and Jones 1983). For this latter case another model has been considered by Königl (1982). He suggested that this string of roughly equidistant HH knots represents a series of Mach disks being formed in an underexpanded jet which is excited to oscillations by the ambient medium.

In the following, we will discuss in more detail another important possibility to form strong shocks (bright knots) in the jets from young stars. As mentioned above, a number of HH objects are apparently located at the end of the jet. The examples known so far are HH 12, HH-DG Tau, HH 33, HH 19, HH 47A, and HH 32A. Within the context of the model proposed above this morphology strongly suggests that these HH objects trace the location of the jets' "working surface". This idea is very much supported by velocity measurements of the jets associated with DG Tau, HH 33/40, HH 46/47 and AS 353A, which indicate that all these flows, as required, are decelerated near or in the bright HH object located at the end of the jet (Mundt and Fried 1983, Mundt et al. 1984, Mundt 1984a). To illustrate this apparent flow deceleration Fig. 8 shows recent results from spectroscopic observations of HH 33. A similar flow deceleration is indicated in the long-slit spectrogram of DG Tau shown in Fig. 7 (for details see Mundt, Brugel, and Bührke 1984).

The schematic structure of a jets "working surface" is shown in Fig. 9 for an adiabatic gas. For a gas with strong radiation cooling a smaller cocoon (or none at all) and a more narrow bow shock will develop. In the model considered here the proper motion of HH objects at the end of the jet would be due to the advancement of the "working

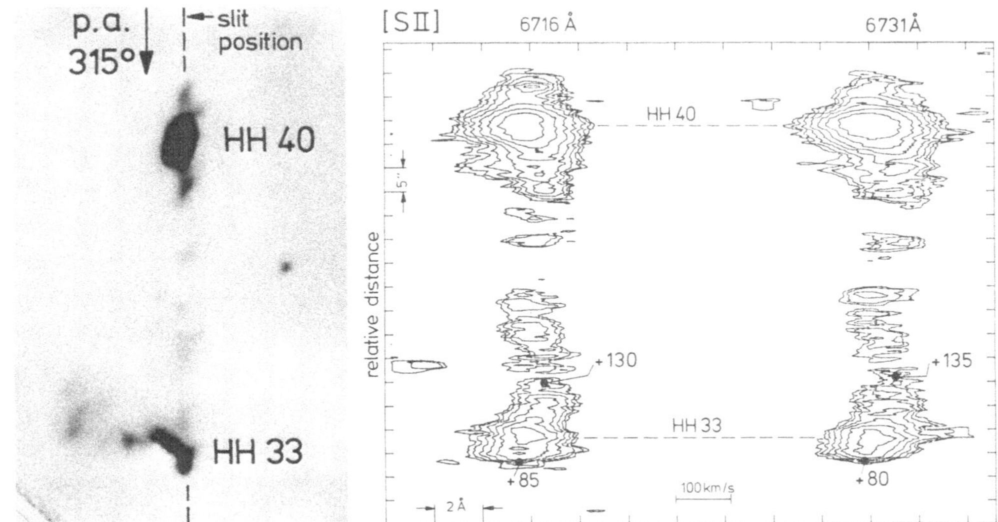

Fig. 8: Left panel: CCD frame of HH 33/40 (figure from Mundt et al. 1984). Right panel: Intensity contour plot of the [SII] lines in a long-slit image tube spectrogram of the HH 33/40 jet. Note the decrease in velocity in HH 33 (figure from Mundt, Brugel, and Bührke 1984).

surface" through the ambient medium (Mundt 1984a). If these HH objects are really tracing the jets' "working surface", both the shock at the end of the beam and the bow shock can in principle contribute to their line emission (see Fig. 9). For V_j=300 km/s and a proper motion of 100 km/s the former shock would have a velocity of 200 km/s and would produce high-excitation lines, while the 100 km/s bow shock would produce lines of considerable lower excitation. A mixture of lines with different degrees of excitation is indeed observed in many HH objects (e.g. Böhm 1983).

As discussed in detail by Mundt (1984a) the HH objects located at the end of the jet would be difficult to understand in terms of the interstellar bullet model. One of his arguments is the flow deceleration in or near the HH objects at the end of the jet discussed above. In the bullet model one should observe the opposite velocity gradients, since ambient matter which has been left behind the bullet after passing its bow shock should have lower velocities (and not higher, as observed).

In summary, it is strongly suggested that some HH objects are powered by jets. This means, they trace the location of high energy dissipation in the strongest shocks occuring in these flows. For those

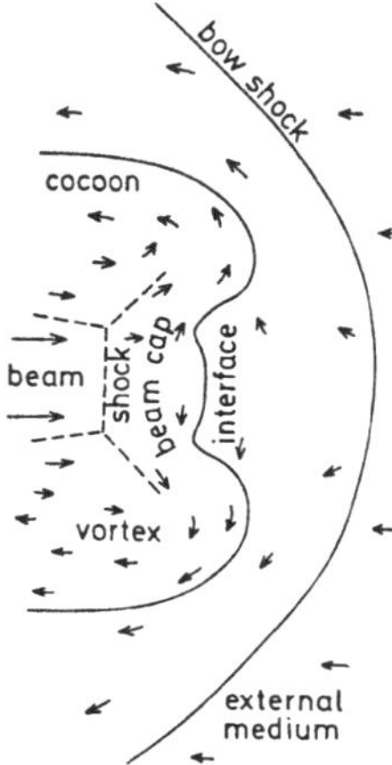

Fig. 9: Schematic picture of the structure of a jets' working surface (figure from Smith et al. 1984).

HH objects located at the end of the jet this idea is supported by the decelaration observed there. The extent to which this model applies to typical HH objects must be tested by future observations. Besides more detailed kinematical studies of known flows, further searches for jets connecting IR sources and HH objects and detailed investigations of such new cases is one promising method to attack these questions.

Acknowledgements:

The author wishes to thank E. Brugel, T. Bührke, H. Kühr, and G. Münch for comments and suggestions or for critically reading the manuscript.

References

Appenzeller, I. Jankovics, I. and Östreicher, R. 1984, **Astron. Astrophys.** 141, 108

Bastian, U., and Mundt, R. 1985, **Astron. Astrophys.** in press

Bertout, C. 1984, **Reports on Progress in Physics,** 47, 111

Bertout, C., Carrasco, L., Mundt, R., and Wolf, B. 1982, **Astron. Astrophys. Suppl.** 47, 419

Bieging, J.H., Cohen, M., and Schwartz, P.R. 1984, **Ap.J.,** in press

Böhm, K.H. 1983, **Rev. Mex. Astr. Ap.** 7, 55

Brown, A., and Mundt, R. 1984, in **Workshop on Stellar Continuum Radio Astronomy,** Boulder, August 1984

Brugel, E.W., Mundt, R., and Bührke, Th. 1984, **Ap.J. (Letters),** in press

Bührke, T., Brugel, E.W., and Mundt, R. 1985, in preparation

Calvet, N., Canto, J., and Rodriguez, L.F. 1983, **Ap.J.** 268, 739

Cohen, M., Harvey, P.M., Schwartz, R.D., and Wilking, B.A. 1984, **Ap.J.** 278, 671

Cohen, M., and Kuhi 1979, **Ap.J. Suppl.** 41, 743
Cohen, M., and Schwartz, R.D. 1979, **Ap.J. (Letters)** 233, L77
Dopita, M.A., Schwartz, R.D., and Evans, I. **Ap.J. (Letters)** 263, L73
Draine, B.T. 1983, **Ap.J.** 270, 519
Drake, S.A., Simon, T., and Linsky, J.L. 1984, in **Workshop on Stellar Continuum Radio Astronomy,** Boulder, August 1984
Dyson, J.E. 1984, preprint
Edwards, S., and Snell, R.L. 1982, **Ap.J.** 261, 151
Edwards, S., and Snell, R.L. 1984, **Ap.J.** 281, 237
Elsässer, H., and Staude, H.J. 1978, **Astron. Astrophys.** 70, L3
Emerson, J.P. et al. 1984, **Ap.J. (Letters)** 278, L49
Feigelson, E.D., and Montmerle, T 1984 **Ap.J. (Letters)** , in press
Finkenzeller, U., and Mundt, R. 1984, **Astron. Astrophys. Suppl.** 55, 109
Gibson, D.M. 1984, in **Workshop on Stellar Continuum Radio Astronomy,** Boulder, August 1984
Graham, J.A., and Elias, J.H. 1983, **Ap.J.** 272, 615
Goldsmith, P.F., Snell, R.L., Hemeon-Heyer, M., and Langer, W.D. 1984, preprint
Hartmann, L. 1982, **Ap.J. (Suppl.)** 48, 109
Hartmann, L. 1984a, **Comments Astrophys.** 10, 97
Hartmann, L. 1984b, **"Cool Stars Stellar Systems and the Sun"** , Eds. S. Baliunas, and L. Hartmann, Lecture Notes in Physics Series 193
Herbig, G.H. 1960, **Ap.J. Suppl.** 4, 337
Herbig, G.H. 1977, **Ap.J.** 214, 747
Herbig, G.H., and Jones, B.F. 1981, **A.J.** 86, 1232
Herbig, G.H., and Jones, B.F. 1983, **A.J.** 88, 1040
Hodapp, K.-W. 1984, **Astron. Astrophys.,** 141, 255
Jones, B.F., and Herbig, G.H. 1982, **A.J.** 87, 1223
Joy, A.H. 1945, **Ap.J.** 102, 168
Kaifu, N. et al. 1984, **Astron. Astrophys.** 134, 7
Knapp, G.R., and Padgett, D.L. 1984, preprint
Königl, A. 1982, **Ap.J.** 261, 115
Kuhi, L.V. 1964, **Ap.J.** 140, 1409
Kuhi, L.V. 1978, in "Protostars and Planets", Ed. T. Gehrels, University of Arizona Press, Tucson, p. 708
Kutner, M.L, Leung, C.M., Machnik, D.E., and Mead, K.N. 1982, **Ap.J. (Letters)** 259, L35
Leverault, R.M. 1984, **Ap.J.** 277, 634
Montmerle, T., Koch-Miramond, L., Falgarone, E., and Grindlay, J.E., 1983, **Ap.J.** 269, 182
Mundt, R. 1984a, in **"Protostars and Planets II",** Eds. D. Black, and M. Matthews, University of Arizona Press, Tucson
Mundt, R. 1984b, **Ap.J.** 280, 749
Mundt, R., Brugel, E.W., and Bührke, Th. 1984, in preparation
Mundt, R., Bührke, Th., Fried, J.W., Neckel, Th., Sarcander, M., and Stocke J. 1984, **Astron. Astrophys.** 140, 17
Mundt, R., and Fried, J.W. 1983, **Ap.J. (Letters)** 274, L83
Mundt, R., and Giampapa, M.S. 1982, **Ap.J.** 256, 156
Mundt, R., and Hartmann, L. 1983, **Ap.J.** 268, 766
Mundt, R., Stocke, J., and Stockman, H.S. 1983, **Ap.J. (Letters),** 265, L71
Norman, M.L., Smarr, L., and Winkler, K.-H. 1984, in **Numerical Astrophysics,** Festschrift in Honor of James R. Wilson, Ed. J. Cantrella
Pudritz, R.E., and Norman, C.A. 1983, **Ap.J.** 274, 677
Rodriguez, L.F., Carral, P., Ho, P.T.P., and Moran, J.M. 1982, **Ap.J.** 260, 635
Sanders, R.H. 1983 **Ap.J.** 266, 73
Sarcander, M., Neckel, Th., and Elsässer, H. 1984, **Ap.J. (Letters),** in press
Schwartz, R.D. 1983. **Ann. Rev. Astron. Astrophys.** 21, 209

Schwartz, R.D., Jones, B.F., and Sirk, M. 1985, **A.J.** , 89, 1735
Smith, M.D., Norman, M.L., Winkler, K.-H., and Smarr, L. 1984, MPA-preprint 150
Snell, R.L., and Edwards, S. 1981, **Ap.J.** 251, 103
Snell, R.L., Loren, R.B., and Plambeck, R.L. 1980, **Ap.J. (Letters)** 239, L17
Stocke, J., Hartigan, P., and Mundt, R. 1984, in preparation
Strom, K.M., Strom, S.E. and Stocke, J. 1983, **Ap.J. (Letters)** 271, L23
Torrelles, J.M., Rodriguez, L.F., Canto, J., Carral, P., Marcaide, J., Moran, J.M., and Ho, P.T.P. 1983, **Ap.J.** 274, 214
Ulrich, R.K., and Knapp, G.R. 1984, preprint
Ushida, Y., and Shibata, K. 1984, **Publ. Astron. Soc. Japan** 36, 105

AN OVERVIEW ON HERBIG-HARO OBJECTS

Jorge Cantó
Instituto de Astronomía, Universidad Nacional Autónoma de México
Apdo. Postal 70-264, 04510 México, D.F., México

I. INTRODUCTION.

The prototypes of the Herbig-Haro nebula class (today known as HH-objects) were discovered by Haro (1950, 1952) and Herbig (1951) during an H_α-emission star survey of the NGC 1999 region in Orion.

The optical emission line spectrum shown by HH-objects is quite peculiar and constitutes the defining criterion of their class. According to Herbig (1969) (see also Schwartz, 1983), the spectra of HH-objects are usually dominated by hydrogen Balmer emission lines and low-excitation emission lines of [O I], [S II], [N I] and [Fe II], together with [O II] and [N II] lines in moderate strength and weak emission of [O III]. In most cases the continuum is absent or rather weak.

The optical visual appearance of HH-objects is also a characteristic of the class since they most frequently appear as small bright condensations or knots surrounded by a more diffuse nebulosity which also shows an HH-object-like spectrum. They can either occur as isolated condensations or in small groups. The linear dimensions of each condensation is of the order of a few hundred astronomical units, and of a few thousands for the whole complex.

HH-objects are always located in or near dark cloud complexes or at least closely related to them; although they only occur in regions where T Tauri stars are also present (Herbig 1962, 1969). These two facts, together with the detection of water masers (Haschick et al. 1983) and IR sources (Strom, Strom and Grasdalen 1975) in their vicinity led to the conclusion that the objects are somehow associated with the process of star formation. This suggestion was first advanced by Ambartzumian (1954).

After several searches for HH-objects in the direction of dark cloud complexes, slightly less than 100 HH-objects have been identified spectroscopically (see Schwartz 1983 and references therein). A good fraction of these objects and their molecular environment have been observed in a wide range of wavelengths, from the X-ray to the radio regime. At the same time several theoretical scenarios have been advanced to explain their peculiar properties.

Most of these observational and theoretical studies have been already summarized in several review papers which have appeared during the last few years. They are: Strom, Strom and Grasdalen (1975), Böhm (1975, 1978a, 1978b), Haro (1976), Cantó (1981) and the most complete and updated one by Schwartz (1983). A deep review of the most relevant aspects of these objects is also available in the proceedings of the

Symposium on HH-objects, T Tauri Stars and Related Phenomena held in Mexico City (February 24-26, 1983) to honor Professor Guillermo Haro. The proceedings appeared in Volume 7 of the Revista Mexicana de Astronomía y Astrofísica (1983, special issue).

The review presented here is intended to give a brief summary of the most recent observational findings and theoretical advances towards the understanding of these objects.

II. SPECTRAL PROPERTIES

1. Line Spectrum

a) Optical

The most remarkable property of the optical spectrum of HH-objects is the large flux (relative to the Balmer lines) of the forbidden and permitted lines of neutral particles and ions of low excitation energy. All known HH-objects show qualitatively the same spectra; quantitative differences range from the so-called "high excitation objects" for which the characteristic lines [O I], [S II], and [N II]) are weaker than Hα and the Balmer decrement is about the recombination theory value, ∿3; to the "low-excitation objects" in which these lines are comparable or even stronger than Hα and the Balmer decrement is higher than 3. Table 1 shows the relative strengths of the most prominent lines in HH-objects 1 and 43, together with the values observed in photoionized regions. HH1 and HH43 are good examples of the "high" and "low excitation" objects, respectively.

TABLE 1. Emission line ratios in HH-objects and photoionized nebulae.

LINES	HH1	HH43	Orion (HII Region)	NGC 7662 (P. Nebula)
[N I] λλ (5198, 5200)	5	160	∿1	<1
[O I] λλ (6300, 6363)	61	630	∿1	<1
[S II] λλ (6717, 6731)	191	1235	24	∿1
Hα	266	720	282	258
Hβ	100	100	100	100

b) Ultraviolet

In the UV only five HH-objects have been observed. Three of them (HH1, 2 and 32) are "high-excitation" objects; their UV spectrum is dominated by lines of high excitation ions such that CIII, CIV, OIII and OIV. The two "low-excitation" objects (HH43 and 47) observed in the UV show spectra totally different from that of the "high-excitation" objects. Here, lines from the Lyman band of H_2, probably excited by fluorescense from atomic hydrogen Lyman α, are the only lines present. See Böhm (1983) and references therein for an extensive discussion on the UV spectra of these objects.

c) Near Infrared.

Nearly 25 HH-objects have been searched for emission in the 2-2.5 μm range. Several of them (HH1, 2, 7-11, 12, 40, 46, 48, 53 and 54) show molecular hydrogen emission lines corresponding to transitions between the first three vibrational states of the ground electron state ($^1\Sigma^+_g$). According to Elias (1980) the observed line intensities exclude any model of UV radiative excitation with lines formed by cascade. This suggests collisional excitation by "warm" (T $\lesssim$ 1000 K) particles as the most likely mechanism. In a few cases the emission appears to extend beyond the optical objects.

2. Continuum Spectrum.

The continuum in HH-objects is in general rather weak. Accurate measurements are only available for eleven objects in the optical and near infrared, five objects in the UV (see Böhm 1983 and references therein) and recently for HH1 and 25 in the FIR (Cohen et al. 1984a, b) and for HH1 and 2 in the radio (Pravdo et al. 1985). The following properties emerge:

i) In general, those objects with the lowest excitation line spectrum exhibit the highest relative blue-optical continuum levels.

ii) All objects show a steep increase of flux towards shorter wavelengths with a smooth transition from optical to UV.

iii) The blue-UV continuum is likely to represent two photon emission from collisionally excited atomic hydrogen since the optical-blue continuum of most HH-objects can be fitted quite well by a free-bound emission plus an enhanced two-photon continuum. Also, the UV spectrum of HH 2H show a peak near 1450 Å, confirming the two-photon nature of the emission.

iv) The infrared-radio continuum in HH1 and 2 appears to be free-free emission produced in the same region than the lines since its strength is, within a factor of two, that expected from the observed H_α intensity.

v) In three cases (HH24, 30 and 101) the red part of the continuum appears to be seriously contaminated by reflected light from a nearby young stellar object. In these objects the continuum is highly polarized (∿10-30%) with the perpendiculars to the polarization vectors pointing towards the young object.

3. Implications and Conclusions from the Spectra.

The interpretation of the spectra of emission nebulae requires a reliable correction for interstellar reddening. For HH-objects, such a correction is commonly performed using Miller's [S II] method, assuming that the intrinsic ratio between the auroral (PD) and transauroral (PS) lines of [S II] corresponds to the collisional excitation value. The following discussion is based on spectra reddening corrected in this way. Recently, however, Münch (1983) has shown that if HH-objects contain well mixed scattering dust, the intrinsic ratio between the [S II] lines can be substantially altered. Consequently, if this effect is not taken into account, the corrected energy distribution in the UV can be overestimated by a large amount (for instance, about one order of magnitude in the case of HH1).

i) Böhm (1983) has summarized several of the basic parameters in HH-objects which can be obtained from the spectroscopic data, without referring directly to the excitation mechanism. They are given in Table 2.

TABLE 2. Basic parameters in HH-objects (Böhm 1983).

PARAMETER	RANGE
Electron Density (cm^{-3})	$2.5x10^3$ - $5.5x10^4$
Electron Temperature (K)	7500 - 12000
Filling Factors	$2x10^{-3}$ - $6.8x10^{-2}$
Ionization Fraction	0.07 0.80
Luminosity ($L_\odot$) (1200 - 11000 Å)	0.1 - 1.4
Ionized Mass ($M_\odot$)	2 - 34
E (B-V)	0.34 - 0.71
A (Hβ)	1.31 - 2.73

We can see that although the electron density and temperature are similar to those found in photoionized regions in molecular clouds, some other parameters are strikingly different. They are the rather small filling factors and the also small ionization fractions. Both point towards a different ionization mechanism that simple photoionization. The other parameter which is of key importance in the interpretation of HH-objects is their visual extinction. This is smaller than the value expected from regions associated with dark molecular clouds, indicating that HH-objects are located in the outer parts of the clouds.

ii) The spectra are similar to that obtained from detailed calculations of the emission of plane parallel shock waves of moderate velocity, in steady state (Raymond 1976, 1979; Dopita 1978; Shull and McKee 1979).

This strongly supports the idea that HH-objects may be represented by the cooling region of interstellar shock waves. However, a detailed comparison between the spectra of the best observed objects, HH1 and HH2H and the models, indicates different shock parameters for different parts of the spectrum. Indeed the optical range suggests a pre-shock density of $\sim$100 cm^{-3}, solar abundances, low pre-shock ionization and a shock velocity in the range 70 - 100 km s^{-1}. In contrast, the high excitation UV spectrum demands a shock velocity of $\sim$200 km s^{-1}, while the 2 - 2.5 μm near infrared emission suggests a velocity of $\sim$15 km s^{-1}. As we will see in Section V.2 this controversy can be resolved if a bow geometry, instead of a plane parallel one, is assumed for the shock. Another spectral feature that plane-parallel shock wave models cannot account for is the steep increase in the continuum flux towards the UV (Section II.2). Models predict a much weaker continuous emission. Dopita, Binette and Schwartz (1982) and Brugel, Shull and Seab (1982 and 1983) have suggested that time-truncated (not fully developed) shock waves can explain this feature.

4. Radial Velocities and Proper Motions.

Consistent with the shock wave interpretation of their spectra, HH-objects tend to exhibit highly supersonic radial and tangential velocities. Both velocities cover a fairly similar range. Fact that favors the interpretation of the observed proper motions as the physical displacement of the emitting material.

The observed radial and tangential velocities show quite interesting properties:

i) Most (∿75%) objects have negative radial velocities. This fact together with the rather low visual extinction exhibited by the objects implies that the emitting material in these nebulae is moving away from the parent cloud.

ii) In some cases, the motions (radial and tangential) are bipolar. That is, different objects or condensations spatially related are moving in opposite directions with respect to what is considered their energy source. Two examples are HH32 (Mundt, Stocke and Stockman, 1983) and HH1-2 (Herbig and Jones 1981).

iii) In almost every case for which proper motions are reliable, the backward extension of the vectors intercepts a pre-main sequence star (either visible or evident as an IR or a radio source).

iv) For HH1, 2, 12, 32, 39, 43, 47 and 50 there are proper motion determinations for individual condensations within each objects (Jones 1983). In all cases, the proper motion vectors of individual condensations within one object are nearly parallel, but with a substantial velocity dispersion among them. These two facts suggest that HH-objects are formed very near the site where we see them and also that the condensations must be short lived (∿50-500 yrs.).

III. THE MOLECULAR ENVIRONMENT.

1. Generalities.

The average H_2 density and temperature of the molecular clouds associated with HH-objects are in the ranges 10^2 - 10^3 cm^{-3} and 8 - 30 K, which are typical of molecular clouds. There is, however, increasing evidence in the sense that HH-objects are closely related to density enhancements ($n(H_2) \gtrsim 5\times10^3$ cm^{-3}; Ho and Barret 1980) within the cloud. This relationship is such that individual objects are not associated with separate density enhancements. Instead, several objects are located around a common density plateau of interstellar dimensions. Also, the position of the density peak is better correlated with nearby IR-sources than with the optical HH-objects. This result is not surprising since the IR-sources are usually extinguished by large amounts (∿20 - 50 mag), while HH-objects have A_V ∿1 mag.

2. Water Masers.

Water masers are frequently found in regions of current star formation, including dark clouds where HH-objects are also present. The required molecular hydrogen densities (∿10^7 - 10^{10} cm^{-3}), kinetic temperatures (∿500 - 100°K) and dimensions of the emitting region (1-100 AU) are those likely to prevail in the envelopes of protostars.

The most extensive survey of HH-objects for H_2O masers has been made by Haschick et al. (1983) and Haschick (1985, private communication). The survey covered an area of 900 square arc-minutes in 20 regions at a sensitivity of ∿10Jy (5σ). Only six masers were detected (four of them were previously known), none of them coincident with the optical object. One is near HH1, three are in the HH7-11 region, one near HH6 and one more is located to the northeast of HH23. These masers are variable, both in intensity and radial velocity, but weaker than those found near young OB stars. Their luminosities are around $10^{-7}L_{\odot}$ which suggest exciting stars with luminosities ∿10^2-$10^3L_{\odot}$ (Genzel and Downes, 1979).

3. High Velocity Molecular Flows.

Molecular gas moving at supersonic velocities is frequently detected in regions of recent star formation. Its presence is revealed through broad emission (or absorption; Mirabel et al. 1985) wings superimposed on the narrow main component emitted by the undisturbed cloud.

Up to now more than 35 regions of high velocity molecular flows are known. They are discussed in detail in Rodríguez et al. (1982), Bally and Lada (1983), Snell (1983, 1984). In all cases the size of the region where the broad wings are detected is much smaller than the surrounding molecular cloud. The gas emitting (absorbing) in the wings is moving highly supersonically and unlikely to be gravitationally bound, as the required central mass is much larger than that observed. Many authors have suggested that these flows represent outflows driven by strong winds from young stellar objects.

More than half (18) of the 35 known high velocity flows are clearly bipolar. That is, blueshifted and redshifted material are located in separate regions, symmetrically located about the presumed energy source. The clearest example is the bipolar outflow associated with IRS 5 in the L1551 dark cloud (Snell, Loren and Plambeck, 1980). Maps of many of these regions are shown in Bally and Lada (1983). Under the stellar wind model, bipolar outflows may be produced either by intrinsically bipolar (Hartmann and McGregor 1982, and Torbett 1984) (or focused very near the star; Snell, Loren and Plambeck 1980) winds; or by an intrinsically isotropic wind which is focused by a large scale agent such as an interstellar dense toroid around the source (see Torrelles et al. 1983; Kaifu et al. 1984 for observational evidence of such structures, and Barral and Cantó (1981) and Königl (1982) for theoretical support).

Seventeen molecular outflows (∿50% of the 35 known) are found in regions containing HH-objects, with 43 HH-objects (∿50% of all known) lying within 10' from the center of one of these outflows (Edwards and Snell, 1983, 1984). In a few cases the HH-objects lie projected within the boundaries of the high-velocity molecular regions, the most remarkable example being HH28, 29 and 102. In most cases, however, although there is a clear association between the molecular outflows and HH-objects, these latter are located far beyond the boundaries of the regions with detectable high velocity CO emission. An outstanding example is R Mon - HH39 (Cantó *et al.* 1982). There the outflow region is ∿2'x3' in size and centered on the stellar object, while HH39 is ∿8' away from the star.

The range covered by the parameters of the molecular outflows associated with HH-objects are given in Table 3 (Edwards and Snell 1983, 1984).

TABLE 3. Basic parameter of the molecular outflows associated with HH-objects. (Edwards and Snell 1983, 1984).

PARAMETER	RANGE
Maximum Velocity (km s^{-1})	4 - 95
Radius (pc)	0.03 - 1.2
Mass ($M_\odot$)	0.003 - 23.5
Momentum ($M_\odot$ km s^{-1})	0.01 - 84.0
Kinetic Energy (ergs)	$4x10^{41}$ - $4x10^{45}$
Time Scale (yrs.)	$1x10^3$ - $5x10^5$

The mean rate of momentum contained in the flows can be estimated by simply dividing its momentum over its dynamical time scale, to yield $\sim 2x10^{-4}$ $M_\odot$, yr^{-1} km s^{-1}. This value represents a lower limit for the rate of momentum delivered by the energy source. A stellar wind of mass loss rate $\sim 10^{-6}$ $M_\odot$ yr^{-1} and terminal velocity ∿200 km s^{-1} fulfills this requirement.

VI. THE EXCITING STARS.

Perhaps the most intringing problem regarding the HH-objects is their energy source. Optical, IR and radio searches indicate that the exciting stars are not within the objects themselves, but displaced by about a light year (or even more). Thus, the identification of the ultimate energy source of HH-objects is not a trivial task.

Figure 1 summarizes the criteria that have been used to isolate the exciting sources. They are:

i) Proximity, that is, the physical involvement of star and nebulosity as in the case of AS 353 A and HH32 (Mundt, Stocke, and Stockman 1983).

ii) Proper Motions, whose backward extension intercepts a young stellar object

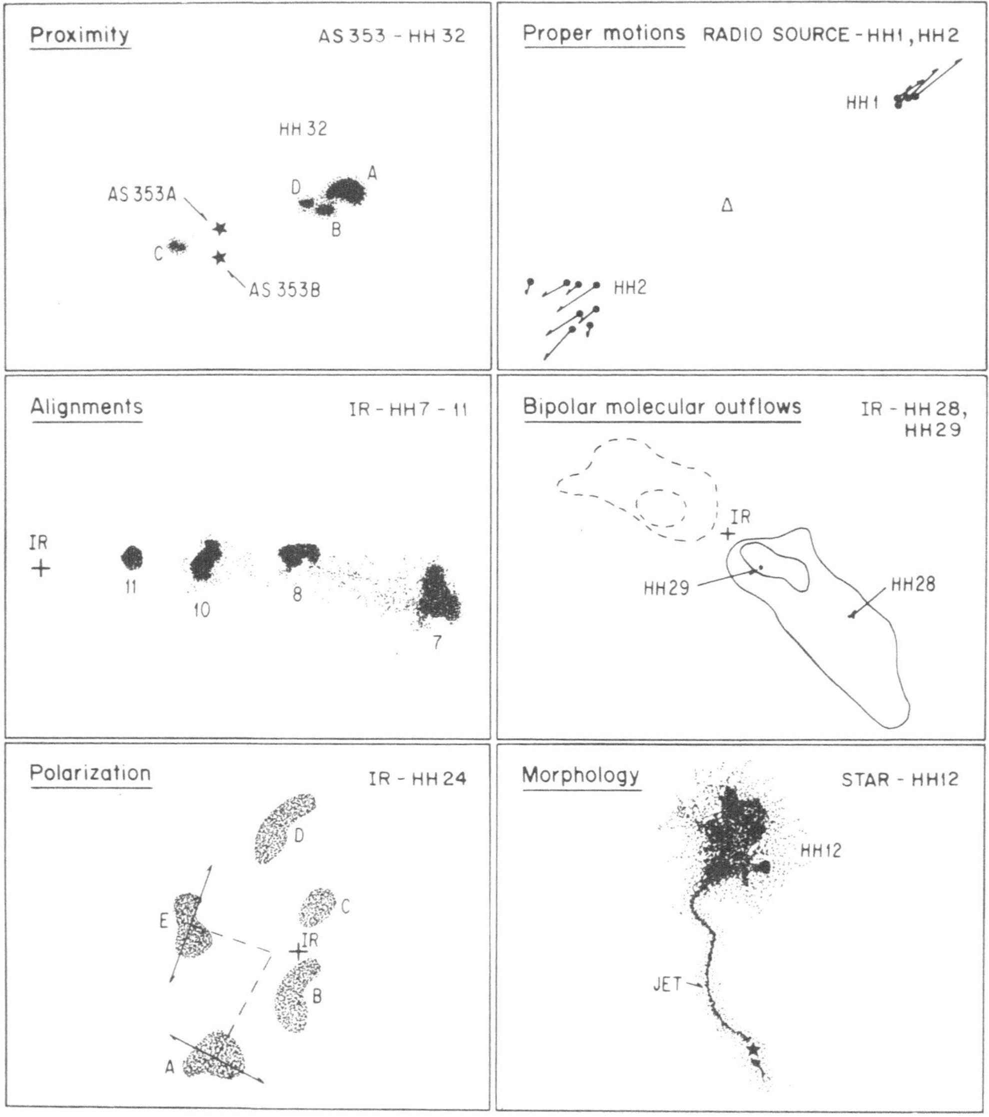

FIGURE 1. Summary of the criteria that have been used to isolate the exciting sources of HH-objects.

(HH1 and 2, and the radio continuum source discovered by Pravdo et al. 1985).

iii) Alignments between HH-objects and the candidate source of energy (HH7-11 and nearby IR source; Strom, Vrba and Strom 1976).

iv) The close relationship between HH-objects and bipolar outflows for which the powering sources are known. The best example is HH28, 29 and IRS 5 in the L1551 region (Snell, Loren and Plambeck, 1980).

v) In highly polarized objects (HH24, 30 and 101) the perpendiculars to the polarization vectors point toward the illuminating source (although it is well established that HH-objects are not reflection nebulae, the fact that in a particular object a substantial part of its continuum derives from reflection of a nearby star, makes this star a good candidate for the excitation source of the object). HH24 is a remarkable example (Strom, Strom and Kinman 1974).

vi) Recent deep photographs of HH-objects in the light of Hα, [N II] and [S II] have revealed the existence of filamentary structures which extend away from the objects (see Mundt 1984 and this volume). Such filaments have HH-like spectra and the emitting material is moving at velocities of several hundred km/s. Typical lengths are ∿0.01 - 0.2 pc, with opening angles of about 3 - 10 degrees. Some authors refer them as "highly collimated jets". These filamentary structures are frequently seen to connect HH-objects with young stars which are thus suggested to be their exciting sources.

In most cases where several of these criteria can be used, they point towards the same candidate. This is the case, for instance, for HH28, 29 and IRS 5 where both the proper motions and the bipolar molecular outflow criteria isolate IRS 5 as the exciting source. In other cases, however, there is a complete disagreement between different criteria. The most remarkable example is HH12 (see Figure 2). There, the proximity criterium favors an IR source which is only ∿30" east of the object (Strom, Grasdalen and Strom 1974). On the other hand, based on the direction of the observed proper motions, Herbig and Jones (1983) suggested star 21 (Table III of that paper). Finally,

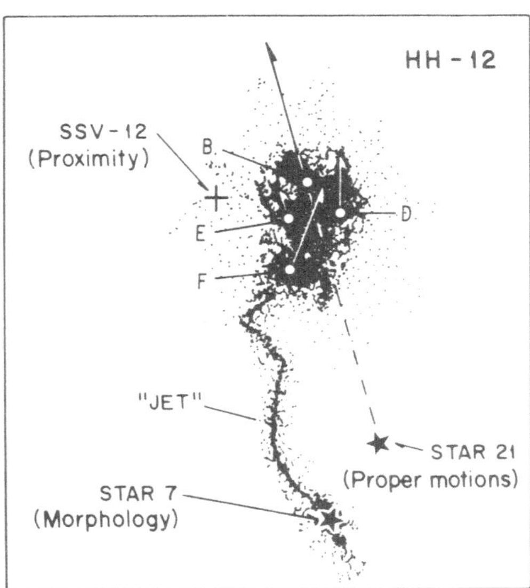

FIGURE 2. The exciting sources of HH 12 according to different criteria.

Strom, Strom and Stocke (1983) prefer star 7 (Table III of Herbig and Jones, 1983) since there is a "jet-like" filament connecting the star and the object.

Another example of the difficulties involved in the identification of the exciting stars of HH-objects is provided by the radio continuum observations of the HH1 - HH2 region. During a near infrared survey of HH-objects, Cohen and Schwartz (1979) detected a bright infrared source near HH1 and along a line joining this object with HH2. This source (which coincides with a visible star) was identified by Cohen and Schwartz as the source of excitation of the objects. This star is usually referred to as the Cohen-Schwartz (CS) star. The identification of the CS star as the energy source of the system appeared to be confirmed when Herbig and Jones (1981) found that the proper motions of the objects indicated a bipolar ejection from a source located in line joining the objects. However, Pravdo et al. (1985) recently discovered a radio continuum source almost exactly centered between HH1 and HH2. Observations of ammonia (Torrelles et al. 1985) and near-IR (Tapia and Roth, private communication) of the region indicates this source as the most likely candidate for the excitation of HH1 and HH2. Figure 3 shows the 6 cm map of the region.

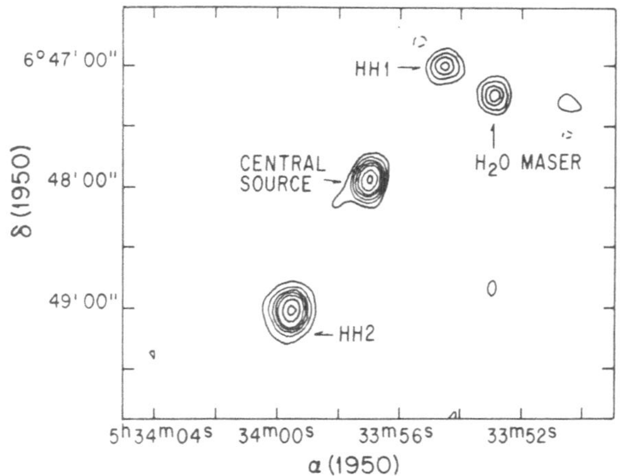

FIGURE 3. The 6 cm radio continuum map of the HH1 - 2 region (Pravdo et al. 1985). Two sources coincide positionally with HH1 and HH2. The source centered in between the objects is interpreted as the energy source of the system. The fourth source coincides positionally with an H_2O maser and is probably excited by an independent star.

Despite these difficulties, the application of the criteria outlined above has resulted in a list of 15 stars which are believed to be the exciting sources of HH-objects (Cohen, 1983). Most of them are invisible in the optical but detected in the IR. Bolometric luminosities, mainly derived from FIR observations, are in the range 5 - 40 $L_{\odot}$ (with the only exception of R Mon), suggesting low mass stars as the energy sources of HH-objects.

V. MODELS.

Current models for HH-objects are based on the assumption that they represent the cooling region of a shock wave. The models can be divided in three categories: those which assume an eruptive or blast event from a central source; those which postulate that the objects are the result of the interaction of a well collimated supersonic jet with the surrounding molecular cloud and those which assume the more or less continuous action of a stellar wind. In the following, I will give a very brief description of each kind of model, with the exception of the jet model which is extensively discussed by R. Mundt in this same volume.

1. The eruptive (or FU Ori) event model (Gyulbudagyan 1975; Dopita 1978; Mundt and Hartmann, 1983) is essentially motivated by the apparent energy problem indicated by the present association of powerful HH-objects with rather low luminosity stars (see below). In this model the HH-objects are thought to be the result of a past FU Ori event undergone by the associated star, at present of low luminosity. It remains to be shown if such an event does in fact produce nebulae with the observed characteristics of HH-objects.

Quite recently, Graham and Frogel (1984) have reported the discovery of a FU Ori-type star on the edge of HH57. Although this finding appears to back up the model, one should remember that the HH-object was already present before the FU Ori event took place. Thus the only firm conclusion we can draw from present evidence is that progenitors of FU Ori-type stars are able to produce HH-objects.

2. The stellar wind models consider that HH-objects (more specifically each condensation within one object) represent bow shocks produced by either the interaction of high-velocity clumps or cloudlets with the ambient molecular material or by the action of a supersonic wind against low velocity dense cloudlets.

A bow geometry for HH-objects is supported by several arguments. First, according to Hartmann and Raymond (1984) a bow geometry can solve the apparent discrepancy between the shock velocities indicated by different parts of the spectrum (see Section II.2). In a bow shock geometry each point in the shock surface has a different shock velocity which is the component of the impinging flow normal to the shock surface. Therefore, the resultant spectrum is a blend of the emissions of a continuum of shocks with different velocities. Second, Choe, Böhm and Solf (1984) have shown that the expected line profiles are quite similar to those observed.

Whether the bow shocks are created by the direct collision of a supersonic wind against dense clumps of cloud material (Schwartz 1975; Schwartz and Dopita 1980) or they are formed as high-velocity clumps or "bullets" plug into the molecular cloud (Norman and Silk, 1979; Rodríguez et al. 1980; Cantó 1983; Tenorio-Tagle and Rozyczka 1984a, b) has been the subject of controversy in the past. Recent evidence tends to favor the bullet interpretation. For instance, the maximum shock velocity derived from the bow shock model of Hartmann and Raymond (1984) for HH2H and HH1 is $\sim$200 km s^{-1}. This velocity is similar to the velocity of the emitting material derived from proper motions (Herbig and Jones, 1981), as expected from the bullet interpretation (see Cantó 1983 for a complete discussion of the correlations expected in either model). Another features favoring the bullet interpretation are the direct correlations between both electron density and the H_α surface brightness (uncorrected for reddening) of the brightest parts of individual condensations in HH2 and the total (radial plus tangential) velocity of the condensations (Cantó and Rodríguez 1985). They are shown in Figure 4.

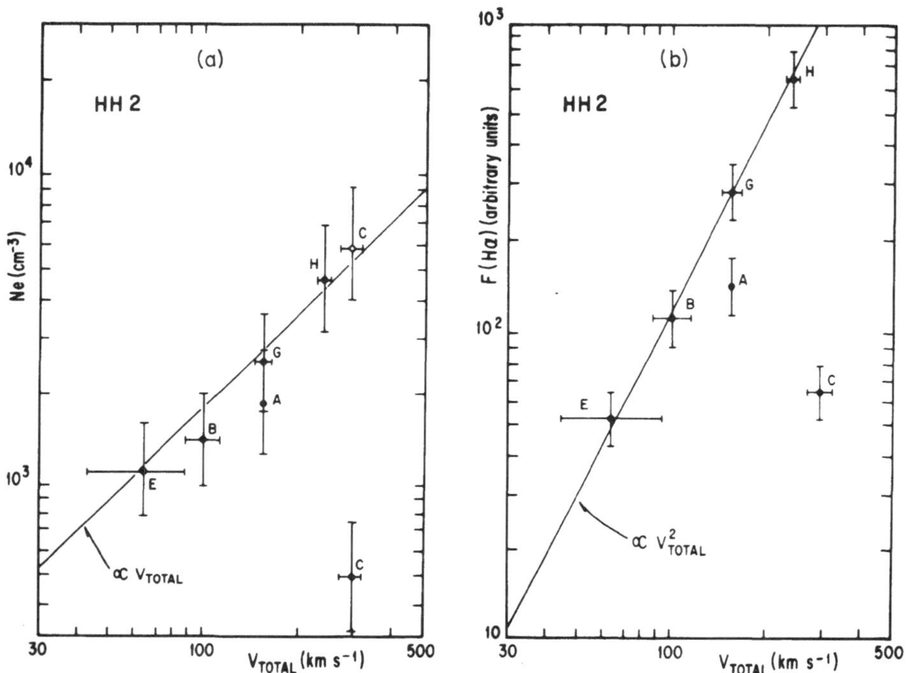

FIGURE 4. Correlations between electron density (a) and H_α surface brightness (b) of the brightest parts of individual condensations in HH2 and the total velocity of the condensations (Cantó and Rodríguez 1985).

These correlations are consistent with a model in which each condensation represents a single shock wave with velocity equal to its spatial velocity, all moving in a same medium (same density and magnetic field) and subject to the same amount of extinction.

Norman and Silk (1979) and Rodríguez et al. (1980) have proposed that the bullets represent dense clumps produced during the breaking-up of the protostellar cocoon (Kahn 1974) which are accelerated to high velocities by the direct action of the stellar wind. Recent hydrodynamic numerical calculation by Sandford and Whitaker (1983) and Tenorio-Tagle and Rozycska (1984) have shown, however, that the transfer of momentum in the wind-cloudlet interaction is rather inefficient. Thus, the acceleration of clumps to high velocities by the action of a stellar wind is unlikely. Instead, Tenorio-Tagle and Rozycska (1984a, b) have proposed that the bullets are the result of a strong shock which overtakes a high density condensation. There, the shock bends around the condensation becoming a conical converging shock. The shocked material condensed upon cooling forms a fast bullet capable of advancing ahead of the shock.

Clearly, if HH-objects are produced by the action of a stellar wind, this wind must be highly anisotropic up to the distances where the optical objects are produced. This is indicated by: the alignment of HH-objects with the presumed exciting sources; the bipolar radial velocities and proper motions; and by the bipolar molecular outflows. At the same time, anisotropy substantially reduces the energy and momentum requirements.

Königl (1982) has proposed a wind channeling model in which an isotropic wind expands into a medium with disk-like density and pressure distributions. As the hot shocked wind bubble expands, it elongates in the directions of decreasing pressure. If the external density distribution can be approximated by a power law in radius with exponent close to 2, the alongated bubble becomes unstable to the formation of a de Laval nozzle. The hot shocked wind is then channeled through the nozzle and accelerated to supersonic velocities. The result is the formation of two oppositely-directed supersonic jets running into the low density environment around the disk-structure. HH-objects represent clumps of matter that have been accelerated by the jet. In this model, the high velocity molecular emission asosciated with HH-objects (Section IV. 3) arises in the shocked ambient material that is swept up by the advancing jet.

Cantó (1980), Cantó and Rodríguez (1981) and Barral and Cantó (1981) have considered the static configuration achieved by an isotropic wind moving into a disk-line molecular environment. The configuration consists of two ovoid cavities filled by unshocked wind and bounded by an oblique shock in the wind (see Figure 5). Since the wind is refracted across the oblique shock, an annular stream of shocked wind is formed. This annular stream slides along the walls of the cavity towards the distant extreme of the configuration. The increasing pressure difference between the inner and outer sides of the cavity forces the flow to turn towards the symmetry axis until it shocks against itself. HH-objects are formed in this second shock.

Although it has to be proved by detailed hydrodynamical calculations, we may consider the possibility that the focused stream is formed, not by a continuous flow, but by a group of clumps of condensed material. They may be either due to thermal or hydrodynamic instabilities in the flow produced upon interaction with the surroundings. The result would be the continuous injection of high velocity dense clumps into the ambient gas; this occurring at a fixed point in the space. The subsequent interaction with the surrounding will finally stop the clumps not far from the injection point. Individual condensations of an HH-object would be represented by luminous bow shocks preceding the high velocity clumps. The high velocity molecular outflows may be produced by viscous coupling between the shocked wind stream and the surrounding molecular cloud (Calvet, Cantó and Rodríguez 1983) while the "jets" associated with HH-objects (Mundt, this volume) may represent the projection, on the plane of the sky, of

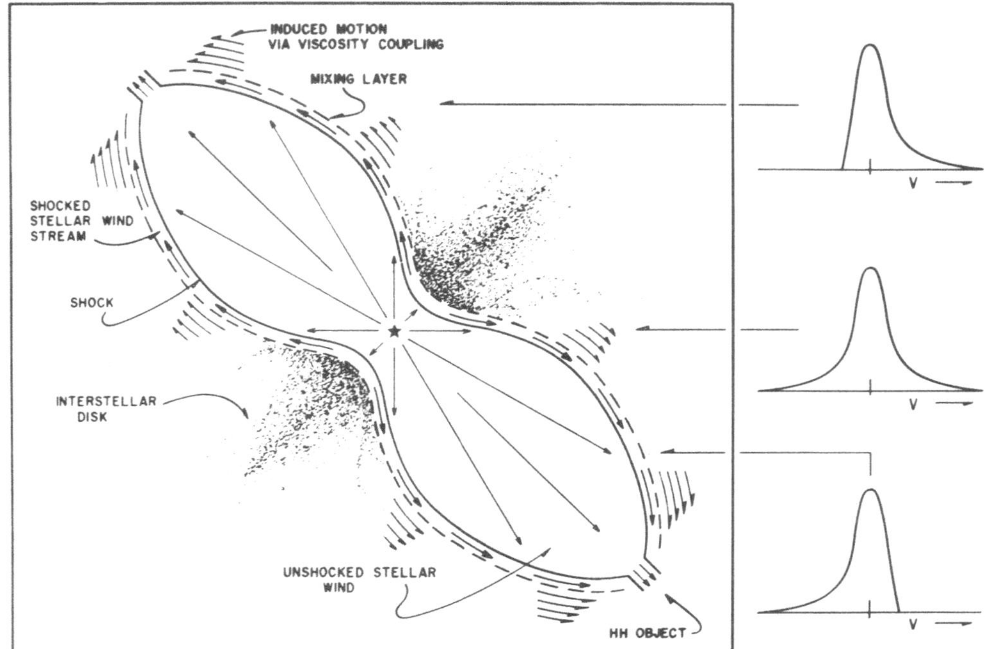

FIGURE 5. Artistic representation of the focusing model for HH-objects. HH-objects are produced on the converging point of the annular stream of shocked stellar wind. The high velocity molecular outflows may be produced by viscous coupling between the shocked wind stream and the surrounding molecular cloud. The "jets" associated with the HH-objects may represent the projection, on the plane of the sky, of the annular stream of shocked stellar wind.

the annular stream of shocked stellar wind. Figure 5 shows an artistic representation of the model. The recent study of Walsh and Malin (1985) of the R Mon-HH39 region seems to confirm the model (see also Cantó and Rodríguez 1981).

The main properties of this model are: 1) its rather high efficiency, since a substantial part of the mechanical energy and momentum in the wind is focused in a small region; 2) HH-objects are always formed at the same point (outside the confining cloud), although the emitting material may have high velocities, and 3) even when the configuration is static, the model is able to explain the high-velocity HH-objects, and the molecular outflows.

VI. CONCLUSIONS.

1) HH-objects are closely related to the star formation process and the early evolution of young stars.

2) Their characteristic line spectrum is produced in the cooling region of bow shock waves, with velocities in the range 40-200 km s^{-1} and pre-shock densities of 50-300 cm^{-3}.

3) The blue-optical continuum is most probably the result of collisionally enhanced two-photon emission, while the infrared-radio emission is due to free-free produced in the same region as the emission lines.

4) The exciting source is not within the optical object but displaced by about ∿0.1 pc, typically, and deeply embedded into the surrounding cloud.

5) A strong stellar wind is the most likely agent by which the embedded source is able to excite the optical object. This wind has to be channeled or focused in a preferential direction(s) in order to explain the small solid angles subtended by the HH-objects with respect to the exciting-source.

6) The exciting sources are most probably pre-main-sequence stars of low mass.

ACKNOWLEDGMENTS

It is a pleasure to thank the organizers of the colloquium on "Nearby Molecular Clouds", 8th E.R.A.M. (IAU) for making this review possible. I also thank L.F. Rodríguez, A. Sarmiento, G. Tenorio-Table, J.M. Torrelles and Professor G. Haro for their helpful comments and suggestions on the manuscript. This work was supported in part by CONACYT (México) grant PCCBBEU-020510. This is contribution N° 152 of the Instituto de Astronomía, UNAM, México

REFERENCES

Ambartzumian, V.A. 1954, Comm. Burakan Obs., N° 13.

Bally, J. and Lada, C.J., 1983, Ap. J., 265, 824.

Barral, J.F., and Cantó, J. 1981, Rev. Mexicana Astron. Astrof., 5, 101.

Böhm, K.H. 1978a, in "Protostars and Planets", ed. T. Gehrels, The University and Arizona Press, 632.

Böhm, K.H. 1978b, in "The Interaction of Variable Stars with their Environments", IAU Coll. N° 42, ed. R.Kippenhahn, J. Rahe, W. Strohmeier, 3.

Böhm, K.H. 1983, Rev. Mexicana Astron. Astrof. 7, 55.

Brugel, E.W., Shull, J.M. and Seab, C.G. 1982, Ap. J. Lett. 262, L35.

Brugel, E.W., Shull, J.M. and Seab, C.G. 1983, Rev. Mexicana Astron. Astrof. 7, 120.

Calvet, N., Cantó, J. and Rodríguez, L.F. 1983, Ap. J., 268, 739.

Cantó, J. 1980, Astr. Ap. 86, 327.

Cantó, J. 1981, in "Investigating the Universe", ed. F.D. Kahn, Dordrecht: Reidel, 95.

Cantó, J. 1983, Rev. Mexicana Astron. Astrof. 7, 109.

Cantó, J. and Rodríguez, L.F. 1981, Ap. J., 239, 982.

Cantó, J., Rodríguez, L.F., Barral, J.F. and Carral, P. 1981, Ap. J. 244, 102.

Cantó, J. and Rodríguez, L.F. 1985, in preparation.

Choe, S.U., Böhm, K.H. and Solf, J. 1984, Ap. J. Lett. (submitted).

Cohen, M. 1983, Rev. Mexicana Astron. Astrof. 7, 241.

Cohen, J. and Schwartz, R.D. 1979, Ap. J. Lett. 233, L77.

Cohen, M., Harvey, P.M., Schwartz, R.D. and Wilking, B.A. 1984a, Ap. J. 278, 671.

Cohen, M., Schwartz, R.D., Harvey, P.M. and Wilking, B.A. 1984b, Ap. J., 281, 250.

Dopita, M.A. 1978, Ap. J. Suppl., 37, 117.

Dopita, M.A., Binette, L., and Schwartz, R.D. 1982, Ap. J. 261, 83.

Edwards, S. and Snell, R.L. 1983, Ap. J., 270, 605.

Edwards, S. and Snell, R.L. 1984, Ap. J., 281, 237.

Elias, J.H. 1980, Ap. J., 241, 728.

Genzel, R., and Downes, D. 1979, Astr. Ap., 72, 234.

Graham, J.A. and Frogel, J.A. 1984, (preprint).

Gyulbudagyan, A.L. 1975, Astrofizika, 11, 511.

Haro, G. 1950, A.J., 55, 72.

Haro, G. 1952, Ap. J., 115, 572.

Haro, G. 1976, Bol. Instituto Tonantzintla, 2, 3.

Hartmann, L. and McGregor, K.B. 1982, Ap. J. 259, 180.

Hartmann, L. and Raymond, J.C. 1984, Ap. J. 276, 560.

Haschick, A.D., Moran, J.M., Rodríguez, L.F. and Ho, P.T.P. 1983, Ap. J. 265, 1981.

Herbig, G. 1951, Ap. J. 113, 697.

Herbig, G. 1962, Advances Astron. Astrophys. 1, 75.

Herbig, G. 1969, in "Non-periodic Phenomena in Variable Stars", ed. L. Detre, Dordrecht: Reidel, 75.

Herbig, G.H. and Jones, B.F. 1981, A.J., 86, 1232.

Herbig, G.H. and Jones, B.F. 1983, A.J., 88, 1040.

Ho, P.T.P. and Barret, A.H. 1980, Ap. J. 237, 38.

Jones, B.F. 1983, Rev. Mexicana Astron. Astrof., 7, 71.

Kahn, F.D. 1974, Astr. Ap., 37, 149.

Kaifu, N., et al. 1984, Astr. Ap. 58, 403.

Königl, A. 1982, Ap. J., 261, 115.

Mirabel, I.F., Rodríguez, L.F., Cantó, J. and Arnal, E.M. 1985, Ap. J. Lett.(submitted).

Münch, G., 1983, Rev. Mexicana Astron. Astrof., 7, 229.

Mundt, R. 1984, in "Protostars and Planets", eds. J. Black and M. Matheus, Tucson, University of Arizona Press, in press.

Mundt, R. and Hartmann, L. 1983, Ap. J. 268, 766.

Mundt, R., Stocke, J., and Stockman, H.S. 1983, Ap. J. Lett., 265, L71.

Norman, C. and Silk, J. 1979, Ap. J., 228, 197.

Pravdo, S.H., Rodríguez, L.F., Curiel, S., Cantó, J., Torrelles, J.M., Becker, R.H. and Sellgren, K., 1985, in preparation.

Raymond, J. 1976, Ph. D. thesis. Univ. Wis. Madison.

Raymond, J. 1979, Ap. J. Suppl., 39, 1.

Rodríguez, L.F., Carral, P., Ho, P.T.P., and Moran, J.M. 1982 Ap. J., 260, 635.

Rodríguez, L.F., Moran, J.M., Ho, P.T.P. and Gottlieb, E.W. 1980, Ap. J., 235, 845.

Sandford, M.T. and Whitaker, R.W. 1983, Mon. Not. R. Astr. Soc., 205, 105.

Schwartz, R.D. 1975, Ap. J., 195, 631.

Schwartz, R.D. 1983, Ann. Rev. Astron. Astrophys., 21, 209.

Schwartz, R.D. and Dopita, M.A. 1980, Ap. J., 236, 543.

Shull, J.M. and McKee, C.F. 1979, Ap. J., 227, 131.

Snell, R.L., 1983, Rev. Mexicana Astron. Astrof., 7, 79.

Snell, R.L., Loren, R.B. and Plambeck, R.L. 1980, Ap. J. Lett., 239, L17.

Strom, S.E., Grasdalen, G.L. and Strom, K.M. 1974, Ap. J., 191, 111.

Strom, S.E., Strom, K.M. and Grasdalen, G.L. 1975, Ann. Rev. Astron. Astrophys., 13, 187.

Strom, K.M., Strom, S.E. and Kinman, T.D. 1974, Ap. J. Lett., 191, L93.

Strom, K.M., Strom, S.E. and Stocke, J. 1983, Ap. J. Lett., 271, L23.

Strom, S.E., Vrba, F.J. and Strom, K.M. 1976, A. J., 81, 314.

Tenorio-Tagle, G. and Rozyczka, M. 1984a, Astr. Ap., 137, 276.

Tenorio-Tagle, G. and Rozyczka, M. 1984b, Astr. Ap., in press.

Torbett, M.V. 1984, Ap. J., 278, 318.

Torrelles, J.M., Cantó, J., Rodríguez, L.F., Ho, P.T.P. and Moran, J.M., 1985, in preparation.

Torrelles, J.M., Rodríguez, L.F., Cantó, J., Corral, P., Marcaide, J., Moran, J.M. and Ho, P.T.P. 1983, Ap. J., 274, 214.

Walsh, J.R. and Malin, D.F. 1985, Mon. Not. R. Astr. Soc., submitted.

THE ENVELOPES OF THE HERBIG Ae/Be STARS

C. Catala

Observatoire de Paris-Meudon
92195 Meudon Principal Cedex, France.

ABSTRACT

First identified by Herbig (1960), the Herbig Ae/Be stars are usually considered as pre-main sequence objects of intermediate mass, and are found at the edges of regions of heavy obscuration.

The study of the interaction of these stars with their surrounding clouds is subordinated to a good knowledge of the structure of their envelopes. The present paper is devoted to a brief review of the observational data available for these stars, and of the constraints on the structure of their envelopes that can be derived from these data.

1 - INTRODUCTION

More than two decades ago, Herbig (1960) proposed to identify a list of 26 objects, which he called "Be and Ae type stars associated with nebulosity", with stars of intermediate mass still in their pre-main sequence (PMS) stage of evolution. This list of 26 Herbig Ae/Be stars had been compiled using the following selection criteria :

(1) The spectral type is A or earlier, in order to eliminate low mass stars or objects too far from the main sequence.

(2) They exhibit emission lines in their spectra : the presence of emission lines was thought to be a characteristic of young objects, by analogy with T Tauri stars.

(3) They lie in an obscured region, because if they are newly born stars, they have not had enough time to escape from their parental clouds.

(4) They illuminate a bright reflection nebula in their immediate vicinity, in order to eliminate those stars that would be projected on molecular clouds.

As Herbig himself mentioned, this list was likely to be incomplete, since it was not the result of a systematic survey. It was then of great interest to try to extend the list of such stars. Recently, a large amount of work has been done in this direction by Finkenzeller and Mundt (1984), who have extended Herbig's original list to 57 candidates.

If Herbig's identification for this group of stars is right, i.e. if they are really PMS stars, their study is of uttermost importance, since they represent a very important range of mass and age in the PMS evolution.

The aim of the present paper is to give a very brief review of the observational material available for these stars and to describe recent progress in the interpretation of some of these observations. In section 2, the problem of the PMS nature of these objects is addressed. A summary of the observational data on these stars is given in section 3. Section 4 gives an example of what can be derived quantitatively from the analysis of high resolution line profiles about the physical structure of the envelopes of these stars. General conclusions are presented in section 5.

2 - PRE-MAIN SEQUENCE NATURE :

The first question that must be addressed about these stars concerns their PMS nature : are they really PMS ? Herbig's criteria are not sufficient to answer this question. Since Herbig's work in 1960, several studies have tried to solve this main problem.

A very important argument in favor of the youth of these stars has been the discovery of a great infrared (IR) excess in most of them (Mendoza, 1966, 1967 ; Geisel, 1970 ; Gillett and Stein, 1971 ; Cohen, 1973, 1975). However, the presence of an IR excess does not constitute direct evidence of the PMS nature of a star, since certain types of stars showing IR excesses are not necessarily young ,like "classical" Be stars, for example, for which the IR excess can be explained by free-free emission alone. In the case of Herbig Ae/Be stars, however, the IR excess has been generally attributed to the presence of hot dust grains in the vicinity of the stars, which argues for the youth of these stars.

A further important argument in favor of the PMS nature of the Herbig Ae/Be stars came from a precise location of these stars in the HR diagram. Strom et al. (1972) have derived the effective temperatures and the surface gravities of 14 members of Herbig's original list, and have concluded that they lie above the main sequence. Later, Cohen and Kuhi (1979) reached the same conclusion by more accurately determining the bolometric luminosities of these stars. Again, this is not a direct proof of the PMS nature of these stars, since objects lying above the main sequence may well be post-main sequence stars.

More recently, Finkenzeller and Jankovics (1984) have compared the radial velocities of 27 Herbig Ae/Be stars from the catalog of Finkenzeller and Mundt (1984) with the radial velocities of the associated molecular clouds. The stellar radial velocities were derived from photospheric lines and the radial velocities of the molecular clouds from observations of the CO, H_2CO, OH and NH_3 molecules (see references in Finkenzeller and Jankovics, 1984). This comparison shows that there is no systematic motion of the stars relative to the molecular clouds. This constitutes a powerful argument for the Herbig Ae/Be stars being physically associated with these clouds, but does not necessarily implies that they are really PMS objects.

In summary, although we have no direct evidence that the Herbig Ae/Be stars are young objects still on their PMS stage of evolution, numerous arguments are in favor of such an interpretation.

3 _ THE OBSERVATION OF THE HERBIG Ae/Be STARS

3-1 Generalities : Figure 1 is a schematic representation of what kind of information can be gained on the different parts of the star's extended atmosphere and its environment by the observations in the different wavelength ranges. In Fig. 1, the stellar atmosphere and its environment have been divided in three parts :

- the gaseous extended envelope, located in the immediate vicinity of the star. Most of the information on this gaseous envelope comes from the line profiles observed in the visible - ultraviolet (VIS-UV) range. A significant fraction of the Herbig Ae/Be stars show evidence for the presence of winds and chromospheres in their gaseous envelopes, as can be derived from the P Cygni profiles and the chromospheric indicators observed in their VIS-UV spectra. X-ray observations can provide information about the presence of corona in these gaseous envelopes, but X-ray emission has been detected only in 3 cases so far (see section 3-4). One could also expect to "see" the cooler parts of the gaseous envelopes thanks to their radio continuum emission, but the latter has never been detected so far in the Herbig Ae/Be stars (see section 3-2).

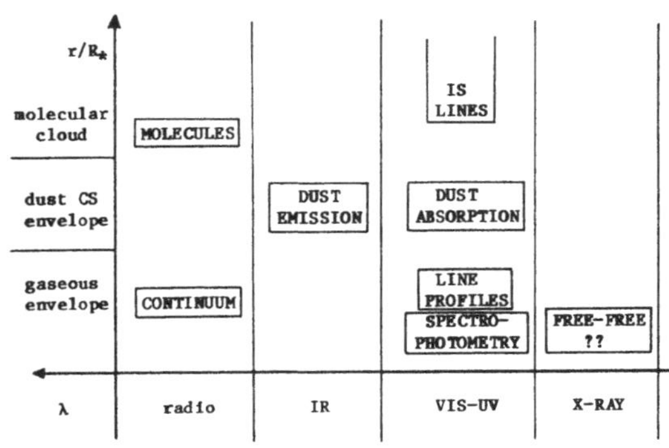

Figure 1: Schematic representation of the possibilities of investigation of the stellar atmosphere and its environment. R_* stands for the stellar photospheric radius

- the dust circumstellar (CS) envelope, containing hot dust grains and located further out from the star. This dust CS envelope can be studied by the observation of dust emission in the IR (section 3.3) and of dust absorption in the VIS-UV range (section 3.5).

- the molecular cloud, in which the star and its CS envelope are embedded. It can be studied by the observation of molecular emission lines in the radio range (section 3.2), and of IS absorption lines, formed within the cloud, in the VIS-UV range (section 3.5).

The following parts of the present section are devoted to a survey of the observational data available for the Herbig Ae/Be stars, in the different wavelength ranges.

3-2 Radio observations : the largest number of observations of the Herbig Ae/Be stars at radio frequencies concern the ^{12}CO and ^{13}CO molecules in the millimetric wavelengths.

Loren et al. (1973) have observed the ^{12}CO ($J=1 \rightarrow 0$) and ^{13}CO ($J=1 \rightarrow 0$) transitions in the direction of 23 Herbig Ae/Be stars, and detected them in all the cases. One of their important results is that in most cases, the maximum of intensity coincides with the position of the star, both in ^{12}CO and ^{13}CO, which constitutes a further proof that these stars are physically associated with the molecular clouds. The fact that the ^{12}CO emission peaks in the direction of the stars in spite of the thickness of the ^{12}CO line is somewhat puzzling. Loren et al. (1973) proposed two possible explanations : the stars lie at the very edge of the molecular clouds, so that the ^{12}CO lines are optically thin, or the clouds are not homogeneous, but clumpy, so that the effective optical thickness of the ^{12}CO line remains small.

Cantó et al. (1984) have surveyed 11 Herbig Ae/Be stars in the ^{12}CO and ^{13}CO lines, trying to detect the presence of high velocity molecular flows. Their results were positive only in two cases. Moreover, the two stars associated with molecular flows have spectral types around B0, i.e. they lie at the hotter boundary of the Herbig Ae/Be class and cannot be considered as typical. The conclusion of Cantó et al. (1984) is that molecular flows are not a characteristic phenomenon for the Herbig Ae/Be stars. The absence of molecular flows associated with these stars raises the question of the physical processes that control molecular flows. Is the absence of molecular flows due to the particular structure of the surrounding material (absence of disk-shaped material for collimating the flows, for example) ? Or is it due to the weakness of the stellar winds of the Herbig Ae/Be stars (insufficient mass loss rates) ? Clearly, if we want to take further steps in the analysis of molecular flows, independent determinations of the mass loss rates of the Herbig Ae/Be stars are required. These independent determinations could come for example from the quantitative interpretation of the line profiles in the VIS-UV range (see section 4).

Several other molecules have been detected in the direction of the Herbig Ae/Be stars : Loren (1977, 1979, 1981) reports the detection of H_2CO, HCN, HCO^+ and CS molecules in the vicinity of several Herbig Ae/Be stars. OH emission has been observed in a few cases, but it is probably not of circumstellar origin (Gahm et al., 1980). No H_2O maser emission has been detected so far (Dickinson, 1976).

Finally, continuous emission at different frequencies has been searched for in several Herbig Ae/Be stars, but the results were negative (Sistla and Hong, 1975 ; Altenhoff et al., 1976; Woodsworth and Hughes, 1977). There remains to be understood what is the basic difference between these stars and radio continuum emitting stars, like P Cygni, that causes this non-detection: difference in mass loss rate, temperature or emitting volume.

3-3 IR Observations : All the Herbig Ae/Be stars show strong IR excesses. In particular, their IR excesses are more intense than those of "classical" Be stars, which constitutes an easy way to distinguish the two classes.

The origin of the excess in the near IR range ($2 \rightarrow 4\ \mu$) is to some extent controversial. Most of the authors attribute it to thermal emission by CS dust (see e.g. Allen, 1973; Cohen, 1973, 1975). In this case, the temperature of the emitting grains should be in the range 1200 - 1500 K. The next step of the analysis consists of finding which grains are likely to be heated to such temperatures by the stellar radiation and to be responsible for the observed IR fluxes. Fitting the data with Plank functions is obviously not sufficient, since usually grains have non-grey absorption coefficients in the IR. On the other hand, several authors have considered the possibility that the near IR excess of some Herbig Ae/Be stars could be due to free-free emission from the H_2^- molecule (Milkey and Dyck, 1973 ; Lorenzetti et al., 1983). In this model, the emission originates from the cooler parts of the gaseous envelope and the electrons are provided by the ionization of metals to their first ionization stage.

Several stars show the 10 μ feature, either in emission (in 2 cases) or in absorption (in 2 cases), attributed to silicate grains (Cohen, 1980). In these particular cases, the presence of hot dust in the immediate vicinity of the star is doubtless.

Very few observations at longer wavelengths are available for the Herbig Ae/Be stars. Harvey et al. (1979) have carried out photometric observations at 40 and 60 μ for 4 Herbig Ae/Be stars. At these wavelengths, these stars still show excesses, which can reasonably be attributed to dust emission. Many other interesting results can be expected from IRAS observations at these wavelengths, but most of these observations are not yet available. However, Wesselius et al. (1984) have already reported IRAS observations of IR excesses in 2 Herbig Ae/Be stars.

Clearly, a big step would be taken if we could determine the nature, the size and the density of the grains in the CS envelopes of the Herbig Ae/Be stars. Such a determination is possible in some particular cases by combining the analysis of the IR observations and of the observations in other wavelength ranges, and has been recently carried out for AB Aur, one of the brightest Herbig Ae/Be stars (Catala, 1983). This kind of approach may provide the beginning of the answer to the difficult problem of the origin of these grains : are they nucleated in the winds of the stars, are they the remnants of the proto-stellar clouds, or is there a permanent infall of dusty material from the surrounding molecular clouds ? Moreover, a good knowledge of the nature, size and density of the CS grains would allow us to address the problem of their dynamics, in particular their dynamical interactions with the stellar winds and with the surrounding molecular clouds.

3-4 X-Ray observations : X-ray observations are essential to understand the physics of the gaseous envelopes of these stars, since X-ray emission can in most cases be attributed to free-free emission from a corona, which can play a major role in the dynamics of the winds, for example. These observations are also very important for understanding the dynamics of the surrounding molecular clouds, since X-ray emission from the stars embedded in the clouds can control the ionization rate of the clouds, and since this ionization rate controls the coupling between the matter and the ambiant magnetic field.

Only 3 stars out of 11 observed have been detected in the X-ray range by the Einstein satellite (Feigelson and de Campli, 1981 ; Pravdo and Marshall, 1981 ; Sanders et al., 1982); therefore, the presence of X-ray emission does not seem to be typical of the Herbig Ae/Be stars. This can be interpreted in two different ways :

i) there are no coronae in the atmospheres of the Herbig Ae/Be stars, or their coronae are geometrically very thin, or their temperature does not reach 10^6 K.

ii) coronae are present in the Herbig Ae/Be stars, but their X-ray emission is reabsorbed by the cooler parts of their gas envelopes, located outwards of the coronae. This interpretation has been presented for T Tauri stars by Walter and Kuhi (1981).

3-5 VIS-UV observations

3-5-1 VIS-UV continuum : The VIS-UV continua of most of the Herbig Ae/Be stars show an anomalous extinction, i.e. an extinction different from that of the "average" IS medium (see e.g. Sitko et al., 1981). This anomalous extinction can probably be attributed to absorption and scattering by the CS dust grains. Sitko (1981) and Thé et al. (1981) have compared for several Herbig Ae/Be stars the energy missing in the VIS-UV range to the energy contained in the IR excess. They have found that these two quantities compare very well in most cases, which is what is expected if the extinction in the VIS-UV and the excess in the IR are both due to a CS dust shell of spherical shape. In some cases, as already mentioned (section 3.3), the analysis of the VIS-UV extinction curve, combined with the analysis of the IR observations, can lead to a determination of the nature, the size and the column density of the CS grains (Catala, 1983).

Garrison (1978) has shown that the Paschen continuum and the Balmer jump of the Herbig Ae/Be stars are well represented by models of classical photospheres surrounded by optically thin expanding envelopes. These models are built on the assumption of optical thinness for the envelopes in the Balmer continuum, which is not supported, and must be considered as preliminary. But the results suggest that, although they are still surrounded by envelopes and embedded in dark clouds, the Herbig Ae/Be stars possess photospheres that look very much like those of main sequence stars.

3-5-2 VIS-UV high resolution spectroscopy: Many IS lines (CaII K, NaI D, KI, MgII, ...) have been observed in the spectra of the Herbig Ae/Be stars. In all cases, the "local standard of rest" velocities of the IS lines are the same as those of the surrounding molecular clouds (Felenbok et al., 1983 ; Finkenzeller and Jankovics, 1984), which supports the claim that these lines are formed within the clouds associated with the stars. Their equivalent widths can therefore be used for a determination of the column densities from the front edges of these clouds to the stars, and of the IS correction that must be applied to the stellar spectra (Felenbok et al., 1983).

Most of the information that we can get on the structure of the gaseous envelopes comes from high-resolution observations of stellar lines, i.e. lines formed within the gaseous envelopes. In that field, the most important work is the one of Finkenzeller and Mundt (1984) who have obtained the profiles of the Hα line, of the HeI 5876 Å line and of the NaI D lines for 57 candidate Herbig Ae/Be stars. One of their main results is that the Herbig Ae/Be stars can be divided into 3 subclasses, depending on their Hα profile : the "double-peak" profile subclass, representing 50 % of the whole class, the "single-peak" profile subclass (25 %) and the " P Cygni" profile subclass (20 %). Figure 2a displays an example of each profile type and figure 2b shows the distribution of the profile types as a function of spectral type. It is reasonable to think that these clear differences in the Hα profiles are due to differences of geometry (spherical or axial, for example) and/or structure (velocity, density, temperature runs) in the gaseous envelopes of these stars. But the interesting question one would like to answer is whether these differences in geometry and/or structure correspond to differences in evolution stage. If the answer to this question was positive, we would be able to classify the Herbig Ae/Be stars as a function of their ages, and we would have a whole sequence of PMS stars at different stages of evolution. Clearly, the first step toward this goal is to understand the geometry and the structure of the envelopes in each subclass. The following section shows an example of a quantitative approach that can be used to derive from the line profiles the physical structure of the regions in which the lines are formed.

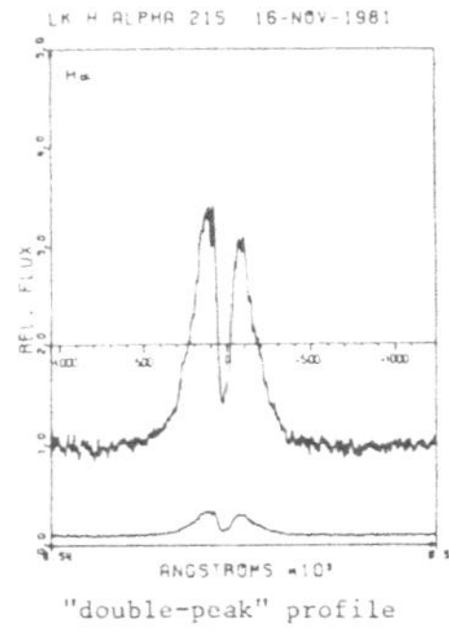

"double-peak" profile

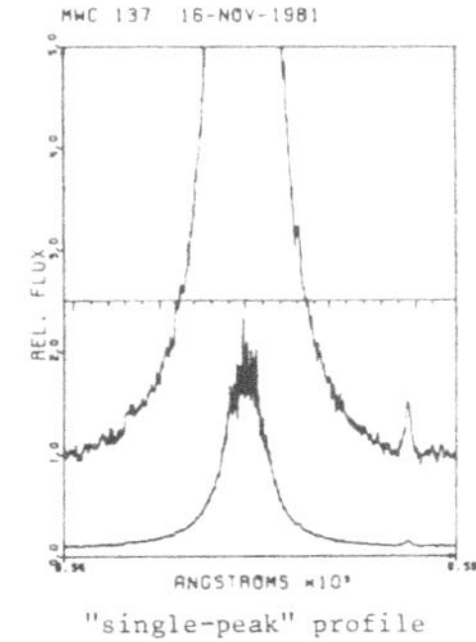

"single-peak" profile

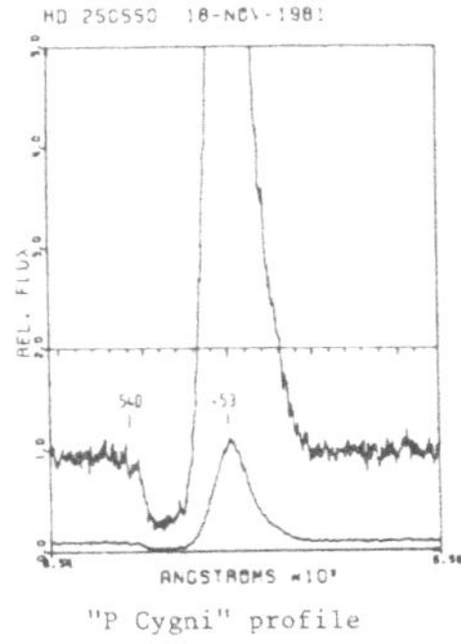

"P Cygni" profile

Figure 2a.(above) Examples of Hα profiles

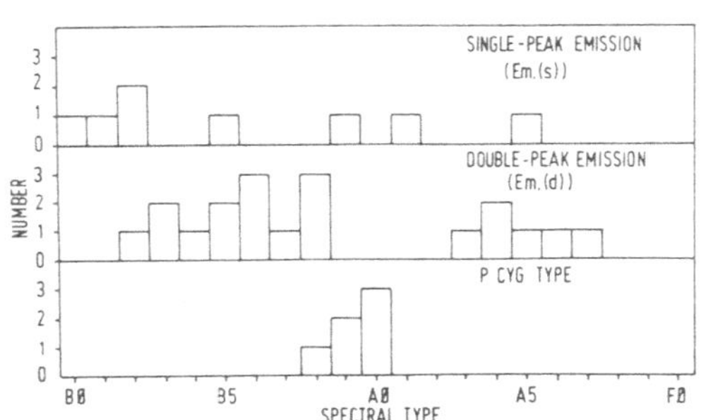

Figure 2b.(left) Distribution of the Hα profile types

(from Finkenzeller and Mundt, 1984)

4 - STUDY OF THE " P CYGNI " SUBCLASS

4-1 Spectral similarities : The first assumption to check is that the Hα profiles are really characteristic of a type of geometry and/or structure. One way of performing this check is to observe as many lines as possible for each subclass and to see if the Herbig Ae/Be stars showing similar Hα profiles also exhibit the same kind of similarities in the other lines. This work has recently been started for the stars of the " P Cygni " subclass, and the results will be published very soon (Felenbok et al., 1985). Figure 3a shows for example the MgII resonance lines for 3 stars of this subclass, and figure 3b, the CaII K line for the same stars. One notes that the profiles have the same shape in the 3 stars. This similarity is also obvious in other lines and for other stars of the " P Cygni " subclass (see Felenbok et al., 1985). The conclusion of this observational study is that the Hα P Cygni profile is indeed characteristic of a given type of geometry and/or structure. This also means that, to study in more detail the "P Cygni " subclass, it is sufficient to study one of its members and to consider it as representative of the subclass. AB Aur, which is the brightest Herbig Ae star in the Northern hemisphere, has been chosen for this more detailed analysis.

To derive quantitative information about the structure of the envelope of AB Aur from its line profiles, a semi-empirical approach has been followed. A brief summary of this semi-empirical approach is given below.

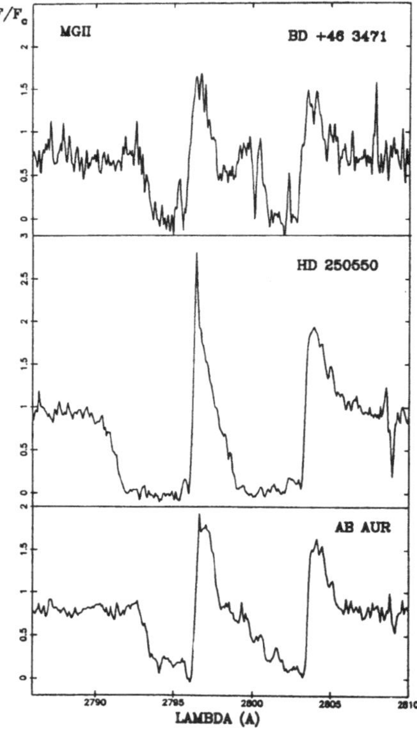

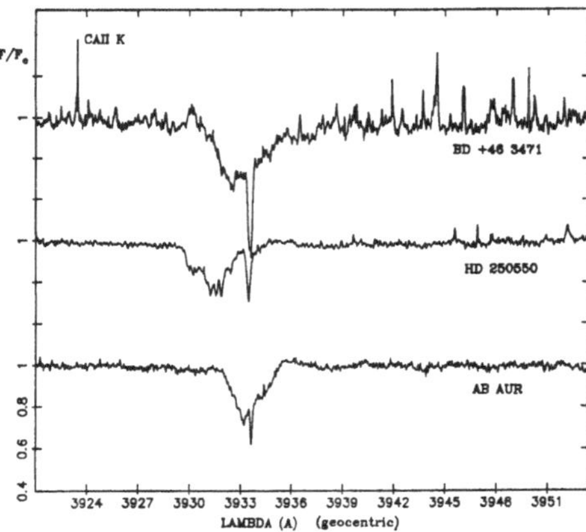

Figure 3a. (left)
The MgII resonance lines of 3 stars of the "P Cygni" subclass, observed with IUE

Figure 3b. (upper right)
The CaII K line for the same stars, observed at the CFH Telescope
(to be published in Felenbok et al.,1985)

4-2 <u>General characteristics of the models</u> : The general type of model that has been considered consists of a spherically-symmetric, expanding envelope surrounding a classical photosphere represented by a model of Kurucz (1979) for $T_{eff} = 10^4$ K and log g = 4. The assumption of spherical symmetry has been made for the sake of simplicity. The presence of an expanding envelope is shown by the P Cygni profiles of Hα (Felenbok et al., 1983) and of the MgII resonance lines (Praderie et al., 1982). The presence of a classical photosphere at the bottom of the wind is suggested by the observation of the Paschen continuum, of the Balmer jump (Garrison, 1978), and of the wings of Hδ (Praderie et al., 1982), which are all well represented by the Kurucz model mentioned above.

The observed maximum blue-shift in the absorption component of the MgII resonance lines on the spectrum presented by Praderie et al. (1982) is V_s=380 km s^{-1}, while neutral sodium D lines show P Cygni profiles with absorption components blue-shifted by 130 km s^{-1} (Felenbok et al., 1983). Since the NaI D lines are necessarily formed at greater distances from the stellar core than the MgII resonance lines, these observations indicate that the wind of AB Aur is decelerated after it has reached its maximum velocity. This maximum velocity is probably reached in the region of formation of the MgII resonance lines.

4.3 Interpretation of the Mg II resonance lines : The next step of this semi-empirical modelling has been a quantitative interpretation of the MgII resonance lines, in order to constrain the dynamical and thermal structure of the envelope (velocity and temperature runs). This work is presented in Catala et al. (1984). A quantitative interpretation of a line profile consists of fitting the observed profile with synthetic profiles computed from models consistent with all the constraints that have been derived from other observations. The problem is that the solution of the fitting is not unique. Several different models can lead to the same computed profile. In other words, it is impossible to calculate unambiguously the model corresponding to the reality by this kind of approach. On the other hand, it is possible to find a whole set of "wrong" models, which means that we are able to derive constraints on the structure of the line formation region. The constraints that have been placed on the structure of the wind of AB Aur by the interpretation of the MgII resonance lines can be summarized as follows :

- a deep, extended and expanding chromosphere is necessary to explain the profile of the MgII lines. The optical depth at 2800 Å of the temperature minimum is 10^{-2} - 10^{-3}. The width of the chromosphere must be higher than one stellar radius. This chromosphere is qualitatively consistent with the observation of the CaII K line (Praderie et al., 1982), of the HeI 5876 Å line and of the CaII IR triplet (Felenbok et al., 1983).

- the mass loss rate of AB Aur must lie between 4.10^{-11} and 7.10^{-9} $M_\odot\ yr^{-1}$.

4-4 Interpretation of the CIV resonance lines : The MgII resonance lines are formed in a very extended region (~ 50 stellar radii), so their interpretation has given only a global vision of the structure of the wind of AB Aur.

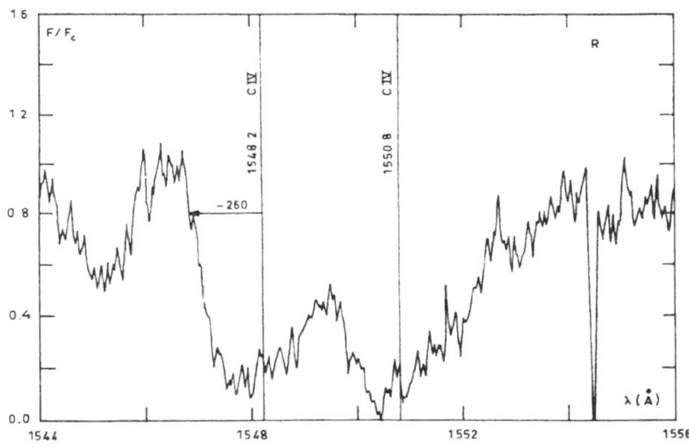

Figure 4. The CIV resonance lines of AB Aur, observed with IUE. The velocity measured on the spectrum is expressed in km.s^{-1}. The "R" indicates the position of a "reseau mark". (from Catala and Talavera, 1984)

The next step of this quantitative analysis had to be the interpretation of lines formed in a more localized area, in order to derive tighter constraints on this area. Figure 4 shows the CIV resonance lines observed in the spectrum of AB Aur. These lines are likely to be formed in the chromosphere (Catala and Talavera, 1984), so we expect to get precise information about the chromosphere by quantitatively interpreting these lines. This work is presently under progress, but preliminary results have already been presented (Catala, 1984). The constraints that have been derived on the structure of the chromosphere are the following :

- the maximum temperature reached in the chromosphere is lower than 19,000 K.
- the width of the chromosphere is lower than 1.5 stellar radii.
- the velocity gradient at the base of the wind must be low (sonic point at more than 1.1 stellar radius from the star's center)
- the velocity near the outer boundary of the chromosphere (i.e. at 2.5 stellar radii from the star's center) must be of the order of 150 $km.s^{-1}$.

Additional modelling is required for a better understanding of the structure of the envelope of AB Aur, and hence of the general structure of the envelopes of the "P Cygni" subclass. In particular, the computation of the hydrogen lines and continua will probably lead to tighter constraints on this structure. Such computations are presently under progress, and will be presented in the near future (Catala and Kunasz, 1985).

4.5 Departures from spherical symmetry : Praderie et al. (1983,1985) have observed with IUE the MgII resonance lines of AB Aur continuously during 40 hours, which corresponds to the presumed rotational period of the star. One of their results is shown in Figure 5, where the maximum blue-shift of the absorption component of the MgII lines (V_s) is plotted as a function of time. The plot clearly suggests a rotational modulation of the MgII resonance lines. The interpretation proposed by Praderie et al. (1983,1985) is the alternation on the line of sight of slow and fast streams, the slow streams coming out from regions of the star where the magnetic loops are closed and the fast ones from regions of open magnetic loops. This model is inspired by observations of the solar wind. We could call it a "solar wind type" model. In the case of AB Aur, since the amplitude of variation of V_s(MgII) is about 100 $km.s^{-1}$, the velocity difference between the fast and the slow streams must be greater than 100 $km.s^{-1}$.

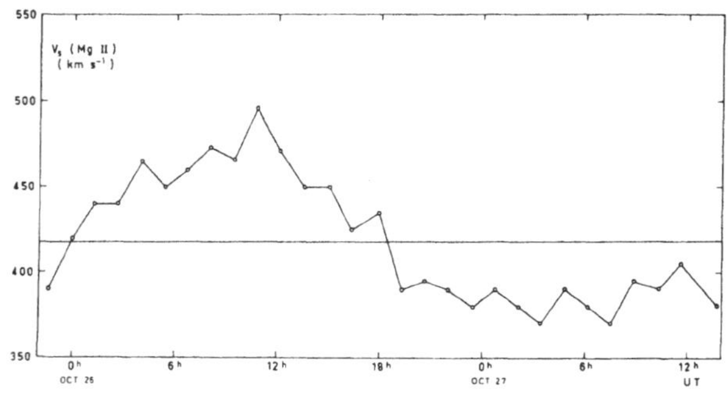

Figure 5. Short term variations of the maximum blue-shift of the MgII k line, V_s(MgII) (from Praderie et al., 1985, by permission).

If this interpretation of the short term spectroscopic variability of AB Aur is right, the assumption of spherical symmetry is obviously no longer valid. One should then ask whether it is useful to continue the computations in spherically symmetric models. My opinion is that this type of calculations is still very useful, since they represent so far our only way of properly understanding the physics of the line formation. Once this physics is well known, it is easy to qualitatively understand the line formation problem in more difficult geometries. Moreover, the computations in spherical symmetry can provide to some extent information on the "average" structure of these envelopes, but in any case, we must be aware that our spherically symmetric models are not that close to the reality, and we must not go too far in the interpretation of the results.

The work on the "P Cygni" subclass that has been briefly summarized here should now be extended to the two other subclasses. First, observational programs should be carried out in order to look for spectroscopic similarities for these subclasses. Second, a theoretical effort should be made to understand the basic structure and geometry of the envelopes of the two other subclasses.

5 - CONCLUSION

From the present review, it is clear that many fundamental questions about the Herbig Ae/Be stars still remain unanswered.

The first set of questions concern the hydrodynamical problems :

- we have almost no idea at present of the mechanism(s) that produce(s) the winds of these stars and that heat(s) their chromospheres.
- it would be of great interest to understand better the interaction between expansion and rotation in these winds. This interaction can have important observational consequences, as we have seen in section 4.5.
- finally, we should address the problem of the interactions between the different components of the stellar environment (gaseous envelope, CS material, surrounding molecular clouds), these interactions including dynamical interaction, ionization, excitation and heating.

The second set of questions are related to evolutionary problems :

- the main problem in this area is probably to understand how these objects become main sequence stars. The Herbig Ae stars, for example, will become main sequence A stars, and will no longer show any wind nor chromosphere. This means that, in addition to finding mechanisms for driving the winds and heating the chromospheres of the Herbig Ae stars, we must also find a way of switching them off as the stars reach the main sequence.
- finally, there is the hope that a more detailed study of the Herbig Ae/Be stars can provide powerful tests for the computed evolutionary tracks. In particular, a comparison between their properties and those of T Tauri stars would provide a good knowledge of the PMS evolution as a function of mass and age.

6 - REFERENCES

Allen, D.A.: 1973, Monthly Notices Roy. Astron. Soc. 161,145

Altenhoff, W.J., Braes, L.L.E., Olnon, F.M., Wendker, H.J.: 1976, Astron. Astrophys. 46,11

Cantó, J., Rodriguez, L.F., Calvet, N., Levreault, R.M.: 1984, Astrophys. J. 282,631

Catala, C.: 1983, Astron. Astrophys. 125,313

Catala, C., Kunasz, P.B., Praderie, F.: 1984, Astron. Astrophys. 134,402

Catala, C.: 1984, Proc. 4th European IUE Conf., Rome, May 1984, ESA SP218, p.227

Catala, C., Talavera, A.: 1984, Astron. Astrophys. 140,421

Catala, C., Kunasz, P.B.: 1985, in preparation

Cohen, M.: 1973, Monthly Notices Roy. Astron. Soc. 161,105

Cohen, M.: 1975, Monthly Notices Roy. Astron. Soc. 173,279

Cohen, M., Kuhi, L.V.: 1979, Astrophys. J. Suppl. 41,743

Cohen, M.: 1980, Monthly Notices Roy. Astron. Soc. 191,499

Dickinson, D.F.: 1976, Astrophys. J. Suppl. 30,259

Feigelson, E.D., de Campli, W.M.: 1981, Astrophys. J. 243,L89

Felenbok, P., Praderie, F., Talavera, A.: 1983, Astron. Astrophys. 128,74

Felenbok, P., Czarny, J., Catala, C., Praderie, F.: 1985, in preparation

Finkenzeller, U., Mundt, R.: 1984, Astron. Astrophys. Suppl. 55,109

Finkenzeller, U., Jankovics, I.: 1984, Astron. Astrophys. Suppl. 57,285

Gahm, G.F., Lindroos, K.P., Sherwood, W.A., Winnberg, A.: 1980, Astron. Astrophys. 83,263

Garrison, L.M.: 1978, Astrophys. J. 224,535

Geisel, S.L.: 1970, Astrophys. J. 161,L105

Gillett, F.C., Stein, W.A.: 1971, Astrophys. J. 164,77

Harvey, P.M., Thronson Jr, H.A., Gatley, I.: 1979, Astrophys. J. 231,115

Herbig, G.H.: 1960, Astrophys. J. Suppl. 4,337

Kurucz, R.L.: 1979, Astrophys. J. Suppl. 40,1

Loren, R.B., Van den Bout, P.A., Davis, J.H.: 1973, Astrophys. J. 185,L67

Loren, R.B.: 1977, Astrophys. J. 218,716

Loren, R.B.: 1979, Astrophys. J. 227,832

Loren, R.B.: 1981, Astron. J. 86,69

Lorenzetti,D., Saraceno, P., Strafella, F.: 1983, Astrophys. J. 264,554

Mendoza, E.E.: 1966, Astrophys. J. 143,1010

Mendoza, E.E.: 1967, Astron. J. 72,816

Milkey, R.W., Dyck, H.M.: 1973, Astrophys. J. 181,833

Praderie, F., Talavera, A., Felenbok, P., Czarny, J., Boesgaard, A.M.: 1982, Astrophys. J. 254,658

Praderie, F., Simon, T., Boesgaard, A.M., Felenbok, P., Catala, C., Czarny, J., Talavera, A., Le Contel, J.M., Morel P., Sareyan, J.P., Valtier, J.C.: 1983, 2nd France-Japan Seminar, Paris, eds. J. C. Pecker and Y. Uchida, p.132

Praderie, F., Simon, T., Boesgaard, A.M.: 1985, in preparation

Pravdo, S.H., Marshall, F.E.: 1982, Astrophys. J. 248,591

Sanders, W. T., Cassinelli, J.P., Anderson, C.M.: 1982, Bull. American Astron. Soc. 14,629

Sistla, G., Hong, S.S.: 1975, Astron. Astrophys. 44,477

Sitko, M.L., Savage, B.D., Meade, M.R.: 1981, Astrophys. J. 246,161

Sitko, M.L.: 1981, Astrophys. J. 247,1024

Strom, S.E., Strom, K.M., Yost, J., Carrasco, L., Grasdalen, G.L.: 1972, Astrophys. J. 173,353

Thé, P.S., Tjin A Djie, H.R.E., Bakker, R., Bastiaansen, P.A., Burger, M., Cassatella, A., Fredga, K., Galin, G., Liseau, R., Smyth, M.J., Viotti, R., Wamsteker, W., Zenge, W.: 1981, Astron. Astrophys. Suppl. 44,451

Walter, F.M., Kuhi, L.V.: 1981, Astrophys. J. 250,254

Wesselius, P.R., Beintema, D.A., Olnon, F.M.: 1984, Astrophys. J. 278,L37

VLA OBSERVATIONS OF POINT SOURCES IN THE RHO OPHIUCHI CLOUD

T. Montmerle[1], Ph. André[1] and E.D. Feigelson[2]

[1] Service d'Astrophysique
Centre d'Etudes Nucléaires de Saclay
91191 Gif-sur-Yvette Cédex, France

[2] Dept of Astronomy
The Pennsylvania State University
University Park, PA 16802, USA

ABSTRACT

Preliminary results of 21cm and 6cm surveys of the Rho Oph cloud with the Very Large Array, in connection with the discovery of 50-70 X-ray sources in the cloud, are given. A first list of radio source identifications is provided; emphasis is put on the variability observed in the peculiar star DoAr 21.

I - INTRODUCTION

a) Activity phenomena above the lower main sequence

It has been known for some time that, at various wavelengths, pre-main sequence low mass stars (including the classical T Tauri stars) display various forms of activity (see, e.g., Montmerle 1984).

One of the most clearcut tracers of activity is X-ray emission. T Tauri stars, but also other (presumed) PMS objects are known to emit X-rays: surveys of nearby molecular clouds with the Einstein Observatory have revealed numerous sources in addition to the classical emission-line objects, increasing the number of PMS candidates by a factor 2 to 3 (see the example of the Chameleon and Rho Oph clouds, e.g. in the review by Feigelson 1984).

In the Rho Oph cloud, short timescale variability on the X-ray sources detected give strong support to the flare interpretation of the emission (Montmerle et al 1983, hereafter MKFG), although a weak underlying corona cannot be excluded.

Since the X-ray activity seems to be quite similar to the solar activity (except for a large scaling factor and a shorter duty cycle), it seemed a logical step to look for non-thermal radio emission from these objects. This is a task for which the NRAO Very Large Array is well suited, because of its high sensitivity.

However, whereas it has been known for some time that T Tauri stars (or, at least, a fraction of them) are radio emitters, the radio flux has been in general attributed to thermal emission from an ionized wind, associated with a large mass loss, $\sim 10^{-9}$ to 10^{-7} $M_{\odot}$ yr^{-1} (e.g. de Campli 1981, Cohen, Bieging, and Schwartz 1982), i.e., $\sim 10^{5}$ to 10^{7} $\dot{M}_{\odot}$.

Hence, a VLA survey of a star-forming region like the Rho Oph cloud has a twofold interest:

- the detected sources (if any) can be unknown objects, or can have counterparts at other wavelengths (IR, X-ray etc.) : in the latter case, one has an additionnal insight into the nature of the emitting stars;
- the observed characteristics of the radio emission give insight into the physical processes witnessing activity phenomena.

b) VLA surveys of the Rho Oph cloud

The Einstein observations of the Rho Oph cloud (using the IPC and HRI instruments) have uncovered 50-70 sources in an area $\sim 2° \times 2°$, with a strong concentration of sources around the densest parts of the cloud (MKFG).

Two of us (E.F. and T.M.) have accordingly conducted surveys of the same area with the VLA at 21cm and 6cm in February 1983, i.e using the C configuration (full details will be published elsewhere).

i) The 6-cm survey was done in the form of 15 fields, (full-width half-power primary beam of 10'), largely overlapping, and centered on the brightest "ROX" sources. The FWHP synthesized beam is ~8". Because of the comparatively small useful field of view, this survey was restricted to the densest part of the cloud.

ii) The 21-cm survey consists of two parts. The first part covers the center region (4 fields) and overlaps the 6-cm survey. The FWHP primary beam is ~30'. The FWHP synthesized beam is ~20". The aim is to get spectral indices of the detected 6-cm sources, no ROX source being more than ~15' away from the primary beam axis (small attenuation). The second part overs 1.5° x 2° of the IPC area, with 12 fields evenly spaced to form a rectangular grid. The aim is to find rapidly possible new sources associated with the cloud (on the basis, e.g., of optical identification).

II - PRELIMINARY RESULTS

a) Looking for cloud sources

In this chapter, we skip all observational details.

Adding 6-cm and 20-cm results, ~ 100 sources have to be analyzed. This includes cloud sources and background extragalactic sources. We note that previous searches for point sources in the Rho Oph cloud (Brown and Zuckerman 1975; Falgarone and Gilmore 1981) yielded a smaller number of sources, because of a lower sensitivity and a smaller area surveyed. There are two steps in determining the association of radio sources with the cloud :

(i) identification with previously known objects, on the basis of celestial position;

(ii) existence of a possible excess in the number of sources over what is statistically expected from source counts. In general, however, this will not yield individual candidates, once step (i) has been performed.

b) Identifications

A number of objects, observed at various wavelengths, are known to be associated with the cloud on the basis of a number of criteria (MKFG; see discussion and refs therein).

The tables give a list of 16 VLA sources, candidates to belong to the cloud with varying degrees of certainty : 9 in the center region (Table 1), 7 in the "streamer" region (end of Table 1 and Table 2) E. of the core. Remarkably - and contrary to the situation prevailing in X-rays - no "bona fide" T Tauri star is detected out of the 11 the cloud contains, but the number of good candidate PMS objects is of the same order.

c) Radio source counts

A number of deep surveys, done recently with the VLA, are now available, over solid angles comparable to our Rho Oph surveys, and down to low to very low intensities ($\ll$ 0.1mJy). The results can therefore be compared in a fairly direct way.

(i) 6-cm survey. The typical rms of our observations is ~0.2 mJy; we take a 5σ upper limit for undetected sources. Above 0.9 mJy, based on the appropriate number-flux density law $\eta(S)$ (Fomalont et al 1984, Bennett et al 1983), we expect to detect 14 sources. We find 16 sources, in addition to those given in Table 1.

This number is thus statistically consistent with all the unidentified sources being extragalactic.

(ii) 21-cm survey. We consider here the "large" survey. This survey can be compared in several ways to the recent surveys by Condon and collaborators (Condon, Condon and Hazard 1982, Condon and Mitchell 1982 and refs therein) : same wavelength, same configuration, ~ same sky area covered. However, our exposures are shorter, hence our sensitivity limit is lower; we also use a smaller number of observations.

Above 2 mJy, we find altogether 69 sources, in addition to sources having identified 6-cm counterparts. Flux density corrections are delicate and often large (see the above references); here, we restrict ourselves to comparing simply numbers of sources actually detected in both surveys.

Between 5 and 150 mJy, Condon et al find 40 sources, whereas we find 41 unidentified sources. If this result is taken at face value, as a preliminary step, the evidence is, like at 6 cm, consistent with all unidentified sources being extragalactic. Because various corrections are necessary, however (in particular because our survey grid is less dense than Condon's) the analysis has to be done in more detail and is in progress.

III - AN EXTREMELY VARIABLE RADIO SOURCE

a) Looking for time variability : the case of DoAr 21

In addition to obtaining spectra, one of the goals of the present work was to look for short timescale variability of the sources, i.e., variability during the allotted observation time of 2x6 hrs, within the 6-cm and 21-cm surveys. This kind of timescale is compatible with the X-ray observations (MKFG). (This is the first time such a systematic study is undertaken for PMS objects.)

A complete study is in progress. However, a first result is that at least one source has been found to be extremely variable in intensity and spectrum.

This source is identified with the star known as Do Ar 21 (Dolidze and Arakelyan 1959). From Table 1, it is seen that it is one of the few radio sources in the survey to be identified optically. It is also seen in the IR and in X-rays, being one of the brightest sources of the cloud at these wavelengths; it is heavily reddened ($A_V = 6$). It is unusual because no emission lines have been visible for the last decade, whereas it exhibited a strong Hα emission in 1949, and a weak one in 1960 (for details, see Feigelson and Montmerle 1985, hereafter FM).

Recent optical spectra show no clear absorption lines, and near-IR photometry suggests a ~G0 spectral type above the main sequence ($L_{bol} \simeq 25\ L_\odot$) (Lada and Wilking 1985).

Therefore, we are dealing with a PMS object that has somehow lost its emission-line (wind?) properties. In what follows, we summarize the forthcoming results on this star (FM). These results are interesting in that they shed doubts on the interpretation of radio emission from all T Tauri stars in terms of mass loss (see also the very recent case of V 410 Tau, Bieging and Cohen 1984). In turn, they are likely to affect current ideas on possible interactions between PMS objects and molecular clouds.

b) Observational results

In addition to the observations of DoAr 21 made in the course of the Rho Oph surveys, and based on early results, VLA runs were obtained in April, June, and September 1983, including 2-cm observations, hence in several configurations (see FM for details).

On February 1983, the source was dramatically brighter than during our other observations, including 17 February. If 18 February is excluded from consideration, the source still exhibits variability at 1.4 and 5 GHz, though at a low level. The spectral index is also clearly changing. The slope of the frequency spectrum was steeply positive, with index + 1.23 ± 0.05 on 18 February, but is flat or negative other times. DoAr 21 exhibited short timescale variability on February 18 : large increase in flux occurred at 5 GHz, from 33.6 ± 0.4 mJy to 48.3 ± 0.3 mJy during 2 ½ hours. The change cannot be attributed to instrumental gain variations since the calibrator 1622-297 was observed during this interval and exhibited random variations of only ± 0.3%. No polarized flux was seen during the event, at a level of a few %.

c) Implications of the variability

The radio emission from late-type PMS stars like DoAr 21 is usually attributed to free-free emission from a spherically symmetric, ionized stellar wind. However, it is shown in FM that the ionized region must likely extend out to $R \simeq 7 \times 10^{14}$ cm with $\dot{M} \simeq 1 \times 10^{-6}\ M_\odot$ /yr. Such a wind model has several serious problems : the mass loss rate exceeds values estimated for even the most extreme T Tauri stars, whereas no Hα emission has been seen on DoAr 21 for many years; assuming a wind velocity $v_\infty \simeq 300$ km/s, the radio emission from such a wind should vary on timescales = $R/v_\infty \simeq 2 \times 10^7$ s, far longer than the observed timescale $\simeq 10^4$ s; and there is no clear way the star can inject energy to accelerate the wind 10^4 radii from its surface.

The principal alternative to wind models is that DoAr 21 undergoes solar-type microwave flares. The spectral index of the 18 February event, +1.2, is similar to 1-10 GHz indices seen on the Sun. The peak observed power, 1.5×10^{18} erg/s/Hz at 5 GHz, is 10^7 greater than that produced by the most powerful solar flares, and the ratio of radio to X-ray emission in DoAr 21 is 10^3 times higher. In fact, the DoAr 21 event most closely resembles radio flares seen in the most active RS CVn stars.

We have made rough estimates of the physical conditions implied by the 18 February event by considering a simple model of (gyro) synchrotron emission from non-thermal electrons (FM). We estimate that the emitting region is $\sim 1 \times 10^{12}$ cm in size, of density $\sim 3 \times 10^{10}$ cm, and pervaded by a magnetic field ~ 500 G. The electron energies are probably more energetic (MeV compared to 10^{-2} MeV) than in solar flares. The duration of the flare, $\sim 10^4$ s or longer, implies that in situ particle acceleration takes place in the loop. Since the observed radio power, spectrum, and risetime are identical to flares occasionally seen in RS CVn system, the physical parameters are likely to be similar as well.

IV - CONCLUSIONS

Although the results described here are not all final, some trends are already emerging, while follow-up VLA observations will soon increase our data base.

a) About 10 to 12 sources seen at 6 cm and 21 cm are linked with the center of the cloud, on the basis of identification with objects known at other wavelengths- except two, which have inverted spectra (as have most radio sources associated with PMS stars). The present evidence is consistent with no unidentified source belonging to the cloud. The conclusion, however, must be taken as tentative in the case of the "large" 21-cm survey.

b) The case of DoAr 21, and of possibly up to 2 other sources, shows that time variability is not unusual. Although less apparent, this trend is reminiscent of the behaviour of the ROX sources (MKFG).

c) The apparent slope of the spectra is itself variable. The spectra are in general inverted, but the slope can be at times decreasing (index = - 0.3 for DoAr 21).

One observational consequence is that, if seen in the direction of a molecular cloud, a radio source with a negative spectral slope is not necessarily an extragalactic source !

d) The probable interpretation of the variability in fluxes and spectra is that PMS objects suffer very intense radio flares ($\sim 10^7$ solar), thus confirming the high level of surface activity traced by the X-rays. This is the first time such flares have been detected in these objects.

One theoretical consequence is that a radio source in a molecular cloud with a positive spectral slope is not necessarily undergoing mass loss, as previously thought. Note, however, that mass loss may be present but undetectable if an other process coexists, like flare generation (in the case of DoAr 21, the radio upper limit to a possible mass loss is still high, being $\sim$a few $10^{-8} M_{\odot}\ yr^{-1}$.)

The presence, followed by the absence, of Hα emission in the optical spectrum of DoAr 21 may show that the mass-loss itself may be variable on timescale of years ("eruptive" mass loss ? activity cycle ?).

This has also possible consequences on the environment of PMS stars, and in particular on the structure of nearby molecular clouds.

REFERENCES

Bennett, C.L., et al., 1983, Nature 301, 686.

Bieging, J.H., and Cohen, M., 1984, in Radio Stars (R.Hjellming and D.Gibson eds) Dordrecht: Reidel, in press.

Brown, R.L., and Zuckerman, B., 1975, Ap.J. (Letters), 202, L125.

Cohen, M., Bieging, J.H., and Schwartz, P.R. 1982, Ap.J., 253, 707.

Condon, J.J., Condon, M.A., and Hazard, C. 1982, A.J. , 87, 739.

Condon, J.J., and Mitchell, K.J. 1982, A.J., 87, 1429.

De Ruiter, H.R., and Willis, A.G. 1977, Astron.Astrophys.Suppl., 28, 211

Dolidze,M.V., and Arakelyan, M.A. 1959, Sov.Astron. A.J., 3, 434.

Falgarone, E., and Gilmore, W. 1981, Astr.Ap., 95, 32.

Feigelson, E.D., 1984, in Cool Stars, Stellar Systems, and the Sun (S.L.Baliunas and L.Hartmann, eds.) Springer-Verlag, p.27.

Feigelson, E.D., and Montmerle, T. 1985, Ap.J. (Letters) in press (FM)

Fomalont, E.B., Kellermann, K.I., Wall, J.V., and Weistrope, D. 1984, Science, 225, 23.

Lada, C.J., and Wilking, B.A. 1985, Ap.J. (in press)

Montmerle, T., Koch-Miramond, L., Falgarone, E., and Grindlay, J.E. 1983, Ap.J., 269, 182 (MKFG)

Montmerle, T. 1984, in Proc. XXV Cospar (Graz), Oxford: Pergamon, in press

Wilking, B.A. and Lada, C.J. 1983, Ap.J., 274, 698 (WL)

TABLE 2. Possible candidates for identification with optical objects within the "streamer" region

ρOph VLA source	Position (1950) α-16h	δ	1.4GHz Flux (mJy)	m_R	Offset (")	Candidates (beam)	Mean number of stars in 20"x20"	Prob.of identi-fication
12	26m47.4s	-24°17'42.8"	6	(16.0)	15	⊡+1(< 30")	0.2	70%
13	26 54.1	-24 56 58.7	6	(15.0)	12	2⊡+2(< 35")	0.2	70%
14	27 00.8	-25 44 04.9	29	(13.0)	20	⊡+3(< 30")	0.6	(**)
15	28 17.8	-23 59 26.5	25	(15.0)	15	2⊡+1(< 30")	0.1	80%
16	28 27.3	-25 02 27.1	27	(17.0)	10	⊡+2(< 25")	0.1	80%

(*) Identifications based on the PSS red print.

(**) Nominally, the procedure used (de Ruiter 1977) diverges when there is more than 0.5 star per beam; we keep this star as a candidate only because it is significantly brighter than the other stars in the error box.

TABLE 1: Candidates for identification within the ρ Oph cloud region

ρOph VLA source	position (1950)		Flux (mJy)		Spectral index	"Oph" source (2)	"ROX" source (3)	IR sources (4)					optical		Comments
	α-16h	δ+24°	1.4GHz	5GHz				E	GSS	VS	VSS	WL	Star	m_R	
1	23m01.6s	16'50"	2→11	2→44	+1.34∓0.3	10	8 (a)	14	23	–	–	:	DoAr21	14.1v	(A)
2	23 32.6	16 45	5→9	9	~ 0.5	4	14 (a')	25	35	–	–	:	⊡	(16.5)	(B)
3	23 42.4	09 49	< 2	1.6	:	–	C9 (b)	–	–	11	–	:	–	–	
4	23 58.5	19 57	34→50	66→90	~ 0.5	6	–	–	:	–	–	:	–	–	(C)
5	24 16.6	22 11	2	3	0.3	12	–	–	:	26	–	4/5	–	–	(D)
6	24 21.5	11 16	<1.5	1.5	:	–	23 (a)	–	:	22	–	:	–	–	
7	24 48.4	19 00	4	:	:	–	–	36	:	14	–	:	⊡	(14.5)	
8	24 50.3	34 10	<2	1.5	:	–	31 (a)	–	:	–	–	:	⊡	(15.0)	(E)
9	25 19.0	19 08	9	29	0.8	–	–	–	:	:	–	:	–	–	(F)
10	28 18.3	23 00	9	:	:	:	–	–	:	:	–	:	⊡ + 6	(9.5)	(G)
11	28 29.0	20 35	3	:	:	:	–	–	:	:	–	:	⊡	(18.0)	(H)
error box diam.	/	/	20"	8"	/	Δα=17" Δδ=110"	(a)=60" (b)=90"	9"~10"		8"~12"			1"	/	/

Symbols used: "–" = not detected; ":" = not available, or not observed; "⊡" = bright star within the radio beam. Magnitudes in parentheses are estimated from the PSS red print (v =variable)

Notes: (1) This work (2) From Falgarone and Gilmore (1981). Note that their sources FG 25 and FG27 are not detected in present survey. Our 3σ upper limits at 1.4 GHz: 2mJy (FG25=14 mJy; FG27= 9 mJy). (3) See MKFG . (4) See references in Table 1 of MKFG; WL= Wilking and Lada (1985).

Comments: (A) Spectral type G0, A_v =5. Detected with the *Einstein* HRI (MKFG). See Feigelson and Montmerle 1985. (B) IR source 1 of Grasdalen (1973); PMS object (see MKFG). (C) Within 1' of ROX C11. Candidate PMS star on basis of radio spectrum only. (D) ROX 22 is 46" away. (E) Spectral type: K7 or M0 (J.Bouvier, private communication). (F) Candidate PMS star on basis of radio spectrum only. (G) Double star (J.Bouvier, private communication), in "streamers". ROX 43 is 44" away. (H) In "streamers". The T Tauri star DoAr 44 (m= 12.5 v) is 42" away, ROX 44 is 55" away.

UPPER LIMITS TO CORONAL EMISSION FROM X-RAY DETECTED T TAURI STARS

M. T. V. T. Lago
Grupo de Matemática Aplicada
Universidade do Porto
Rua das Taipas 135
4000 Porto, Portugal
M. V. Penston
Royal Greenwich Observatory
Herstmonceux Castle, Hailsham
Hailsham, East Sussex BN27 1RP,UK
R. M. Johnstone
Astronomy Centre
University of Sussex
Falmer, Brighton, BN1 9QH,UK

Summary: We have set a new very low limit to emission in coronal lines in three T Tauri stars - two of which are detected as X-ray sources. This non-detection is surprising, at least in the case of GW Orionis, where, by analogy with the Sun, one could rather have expected to find the line at that level. We discuss the consequences of this result and outline a current attempt to measure the magnetic field in T Tauri stars.

Before addressing the problem of coronal line emission in T Tauri stars it is useful to begin by summarizing a model that has been proposed for the T Tauri star RU Lupi. In this model the presence of a magnetic field and linearly polarized Alfvén waves propagating outwards constitute the primary mechanism for driving the wind. Both theory and observations constrain the wind solutions (Lago 1984) and the overall picture is as follows:

- the wind velocity starts with very low values near the base of the chromosphere but (due to the presence of the waves) accelerates very rapidly reaching velocities of order of 240 Km s^{-1} quite close to the stellar surface. In order to explain the observed stratification in the widths of the lines (Lago 1979), dissipation of the waves is assumed to occur before the flow reaches the escape velocity. Therefore the wind will decelerate afterwards due to the gravitational forces. The higher excitation lines would be produced in this decelerating region. The IUE high resolution observations have confirmed such expectations since C IV, Si IV, Si III] and C III] are observed to be narrower than the strong Mg II, Ca II and Balmer lines (Lago, Penston & Johnstone 1984a). Fig. 1 summarizes the observational constraints on the wind of RU Lupi (Penston & Lago 1984).

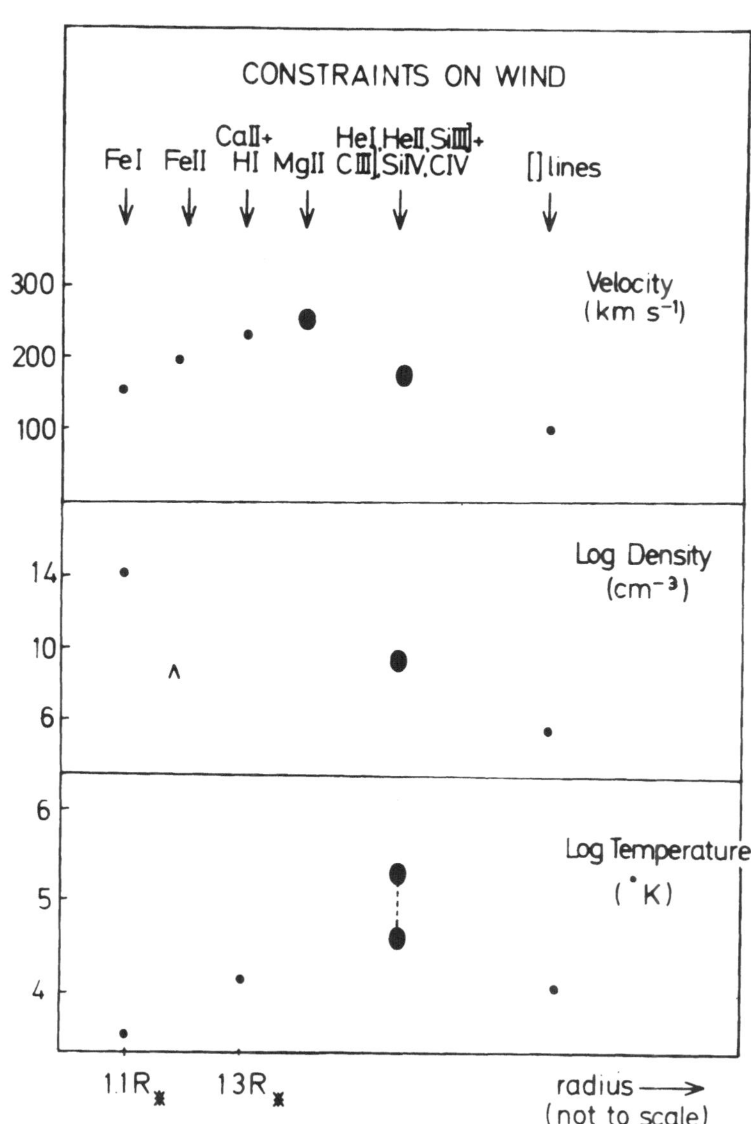

Fig. 1 - Observational constraints on the wind of the T Tauri star RU Lupi.

This model is able to reproduce the velocity - distance relationship suggested by the optical line profiles and also the rather restricted density requirements imposed by the observations if a key parameter, the mean magnetic field, is in the range 600 - 800 G. It seems able to reproduce as well the line profiles observed for the hydrogen lines used to test the model. Furthermore, it suggests a possible explanation for the variation of the temperature through the line emitting region: the heating occurs as a result of the wave energy dissipation (over a short range of distances from the star surface) due to the density gradient (Belcher 1971).

What about the regions where the temperature is even higher? From low resolution UV spectra of over 20 T Tauri stars "the hottest" line observed is N V and even this line seems to be present (and very weak) in only a few stars (Lago, Penston & Johnstone 1984a). In the optical, the forbidden coronal lines of [Fe X] λ6375 Å and [Fe XIV] λ5303 Å , indicative of temperatures up to 2 10^6 K have also been sought in several T Tauri stars but there are no detections so far. In T Tau and GW Ori we set an unprecedentedly low limit on the equivalent width of 20 mÅ to any λ6374 Å [Fe X] emission. Fig. 2 (Lago, Penston & Johnstone 1984b) plots the ratio of stellar to solar surface fluxes as a function of temperature (for temperatures ranging from below 10^4 to over 10^7 K) for RU Lupi and two other T Tauri stars that have been detected as X-ray sources, GW Orionis and T Tauri.

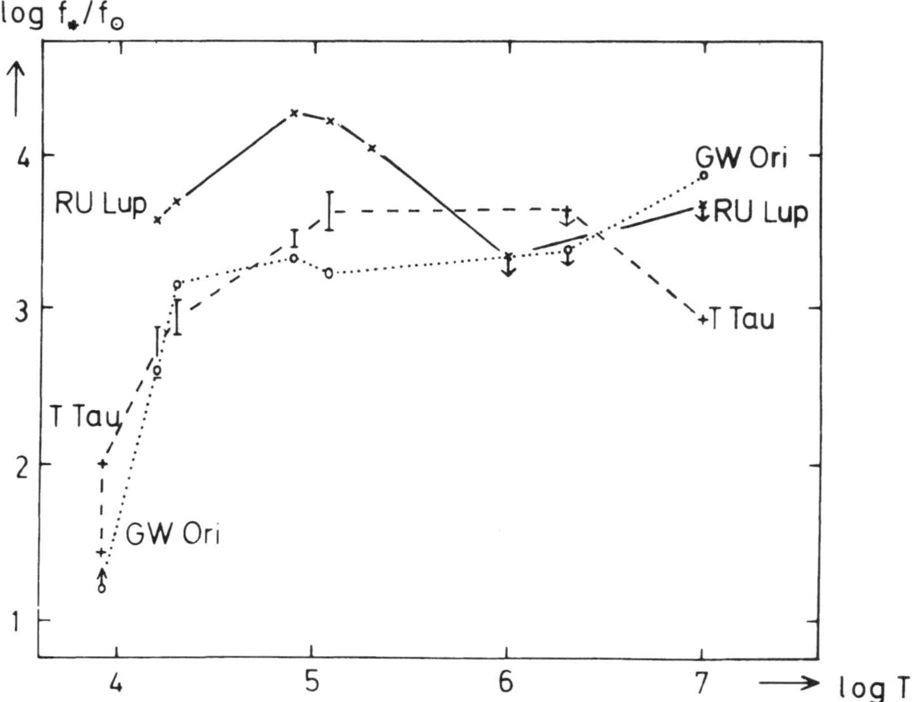

Fig. 2 - Plots of the ratio of stellar to solar fluxes in three T Tauri stars.

Several conclusions may be drawn:

- the surface fluxes from chromospheres, transition regions and coronae are much higher in T Tauri stars than in the Sun, and furthermore, the proportion of material at the various temperatures is also different;

- there are structural differences between the T Tauri stars themselves:

in GW Orionis the line fluxes emitted by gas at different temperatures in the chromosphere, transition region and corona, compared to the same regions in the Sun, increase with temperature; the upper limit observed for the coronal line flux lies slightly below a smooth curve fitting the ultraviolet and X-ray data. Possible explanations for such a failure to detect the coronal emission at the level that might be anticipated from the strengths of the transition region lines will be discussed later; in T Tauri there seems to be a peak in luminosity (relative to the Sun) at 10^5 K; in this case the limit to the coronal line flux lies close to a smooth curve fitting the other data; it is however remarkable how different the X-ray flux ratio is from that in GW Ori, given the similarity of the two stars at lower temperatures; in RU Lupi there is a clear difference from the other stars: the emission from the chromosphere and transition region is much stronger than in the Sun (and appreciably stronger than in the other T Tauri stars) but there is no corona detected at the corresponding level in either coronal lines or X-rays. The weakness of the X-rays has been suggested to be due to absorption (Walter & Kuhi 1981) but this is certainly not the explanation for the weakness of the forbidden line of [Fe X] . There are several possible explanations:

- a high density, $Ne > 10^{10}$ cm^{-3} in the [Fe X] and [Fe XIV] emission regions might lead to collisional deexcitation of the coronal lines - the true emission measures would then lie higher than the limits on effective emission measures plotted in Fig. 2. In RU Lupi the C III]/Si III] ratio suggests a density 8 10^9 cm^{-3} at T = 5 10^4 K and it seems unlikely that the pressure in the [Fe XIV] emitting region would be so much higher than in the chromosphere and transition region. Such a high pressure would need to be confined by a magnetic field and such confining fields are very likely to be unstable (Mullan & Steinolfson 1983);

- coronal line and/or X-ray emission which are observed at different times might be variable (maybe there are flare-like outbursts such as those reported for the pre-main-sequence stars in the Rho Ophiuchi cloud (Montmerle et al 1983));

- the X-ray mechanism could be non-thermal, perhaps associated with some particle acceleration mechanism - magnetic field reconnection, for example.

How can we try to explain the differences in behaviour displayed in Fig. 2 for the three T Tauri stars? Earlier in the session, while reviewing some of the models that have been proposed for T Tauri stars, Bertout mentioned the difficulties for successful profile fitting and commented on the nonexistence of a satisfactory model. We have a different opinion and would like to caution you against the use of models of the "average" T Tauri star, a concept rather difficult to maintain in the face of the observational evidence. Fig. 3 displays the UV spectra of two T Tauri stars with similar optical spectra (and belonging to the same emission class according to the Herbig and Rao catalogue). As can be seen, the UV spectra are not at all alike and there are many such contradictory examples.

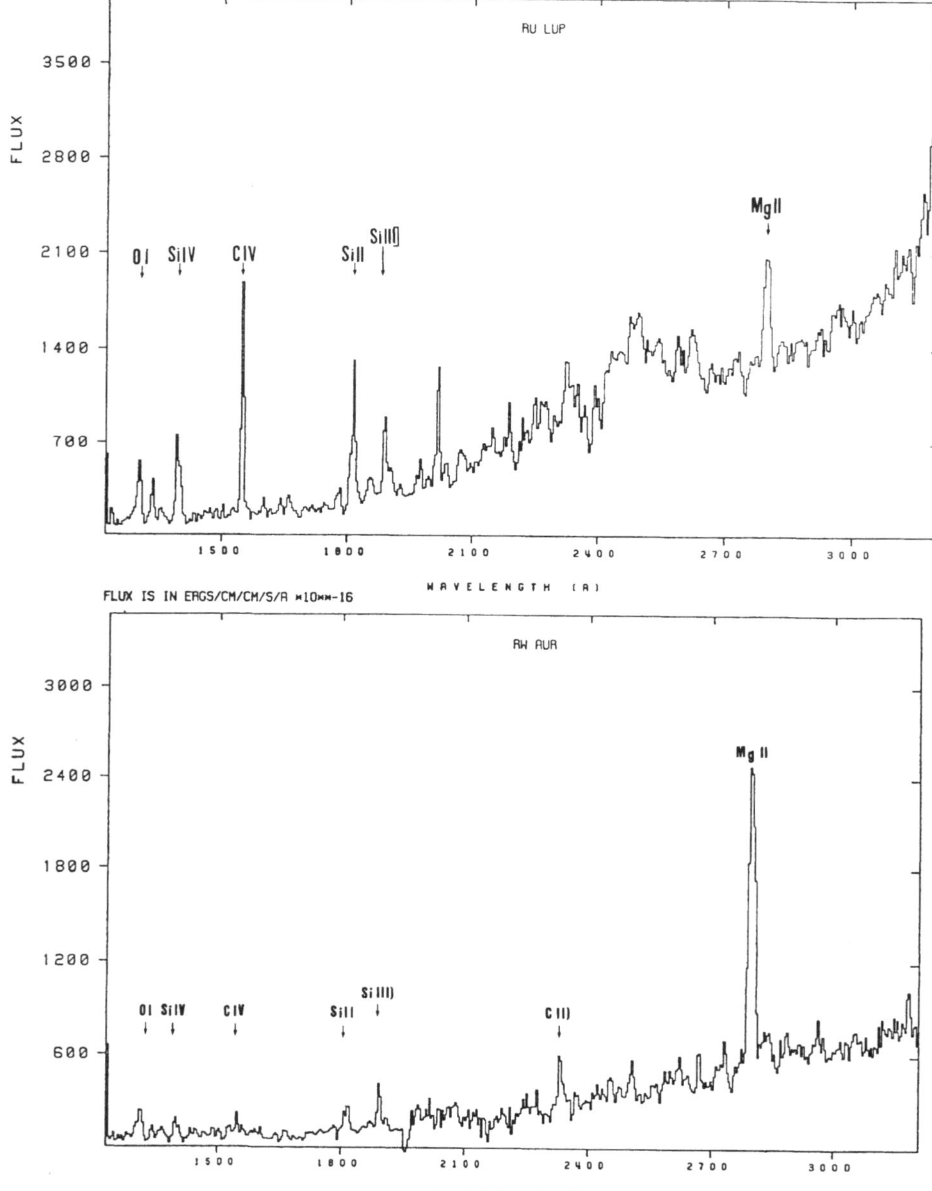

Fig. 3 - Low dispersion IUE spectrum of two T Tauri stars belonging to the same emission class. (Mg II is saturated in both spectra and Si II λ1812 Å is also saturated in the spectrum of RU Lupi).

It is now necessary, while attempting detailed model and profile fitting, to assume that T Tauri stars should be dealt with as individuals, very different one from the other. The available high resolution data are now also much too good for average models to have much meaning. This leads us to the final part of this talk: why do T Tauri stars differ so much one from the other? A possible explanation is suggested by the model we described earlier on; the free parameters in this model are: the temperature, density, magnetic field intensity and magnetic field perturbation (wave amplitude) at the reference level, roughly the base of the chromosphere. We suspect that what one observes in the stellar spectra will mainly depend on the intensity of the magnetic field and the distance at which the dissipation of the wave energy occurs: the closer to the star the dissipation, the higher are the temperatures that will be reached with less mechanical energy available to accelerate the wind. This will have implications on the presence (or absence) of a corona around the T Tauri stars as well as on the interaction of the wind with the circumstellar material.

We have recently made an attempt to measure the magnetic field in T Tauri stars. The Royal Greenwich Observatory spectrograph and the new Pockels Cell Polarimeter were used at the Anglo Australian Telescope to obtain simultaneous spectra in both left and right hand circularly polarized light and to search for a wavelength shift due to Zeeman splitting. The results are still very preliminary and in particular the errors are still uncertain. They may however indicate somewhat different magnetic fields for the two stars:

$B \simeq 530 \pm 220$ G for RU Lupi

$B \simeq 230 \pm 280$ G for GW Orionis.

These results must still be treated very cautiously, but if confirmed, either by repeated observations or by a final version of the analysis of the present data they would certainly be very encouraging for the model described earlier. The value found for the magnetic field in RU Lupi agrees well with the range of values predicted by the model of the wind in this star which fits the line-width observations. Further theoretical work (in progress) will also allow the extension of this model to other T Tauri stars and to give estimates of the range of values of the magnetic field intensity required by the observational constraints on other T Tauri stars.

Acknowledgments: M.T.V.T.L. acknowledges partial finantial support from Fundação Gulbenkian, from the Instituto Nacional de Investigação Científica and from the Symposium's Organizing Committee. We thank the British Council for travel funds which allow the authors to complete this paper.

References:

Belcher, J. W., 1971,Astrophys. J., 168, 509.
Lago, M. T. V. T., 1979, DPhil. thesis, University of Sussex, Brighton.
Lago, M. T. V. T., Penston, M. V. & Johnstone, R. M., 1984a, Fourth European IUE Conference, ESA- SP 218, 233.
Lago, M. T. V. T., 1984, Mon. Not. R. astr. Soc., 210, 323.
Lago, M. T. V. T., Penston, M. V. & Johnstone, R. M., 1984b, Mon. Not. R. astr. Soc., 211, in press.
Montmerle, T., Koch-Miramond, L., Falgarone, E. & Grindlay, J. E., 1983, Astrophys. J., 269, 182.
Mullan, D. J. & Steinolfson, R. S., 1983, Astrophys. J., 266, 823.
Penston, M. V. & Lago, M. T. V. T., 1982, Third European IUE Confrence, ESA-SP 176,95.
Walter, F. & Kuhi, L. V., 1981, Astrophys. J., 250, 254.

ROTATION AND X-RAY ACTIVITY IN T TAURI STARS

J. Bouvier and C. Bertout
Institut d'Astrophysique, 98 bis, bld Arago
F-75014 PARIS

W. Benz and M. Mayor
Observatoire de Genêve
CH-1290 SAUVERNY

I. Introduction

Although T Tauri stars have been studied extensively since their 1945 discovery by Joy, they remain puzzling in many respects. These young low-mass stars, which are still contracting toward the main-sequence, display large and often irregular light variations at all wavelengths, and their spectra reveal many emission lines (H, CaII, FeI, FeII ...) usually superimposed on a normal late-type absorption spectrum. These peculiarities are generally interpreted in terms of photospheric and chromospheric activity, whose relationship to stellar magnetic fiels is well-documented for main-sequence stars. More direct evidence of magnetic activity is offered by the periodic light variations observed in some moderately active T Tauri stars, which can be interpreted best as a rotational modulation by large surface dark spots (Rydgren et al., 1984).

Studies of activity due to magnetic fields in late-type main-sequence stars have shown the existence of relationships between different activity criteria (e.g. CaII emission, X-ray flux) and rotational velocity (e.g. Noyes et al., 1984 ; Pallavicini et al., 1981). These results were expected from theoretical work on the dynamo process, which arises from the interaction of differential rotation with the deep convective envelope present in late-type stars. Dynamo models predict that the strength of the surface magnetic field increases both with rotational velocity and with the depth of the convective zone. The surface magnetic field in turn is believed to be responsible for activity in late-type stars.

Whether activity of T Tauri stars can also be attributed to dynamo processes is a question that has not been addressed yet because of the scarcity of data on rotation in T Tauri stars. The only large-scale study of rotation in pre-main-sequence objects so far (Vogel and Kuhi,

1981) was hampered by detector limitations that did not allow them to detect rotation velocities smaller than 25-35 km/sec. Although Vogel and Kuhi discovered that, contrary to expectations, the rotation velocity of most T Tauri stars was rather small, they could only give upper limits for the rotation rate, making it impossible to study any possible relationship between rotation and activity.

In this paper, we report measurements of rotational velocities for a number of T Tauri stars. The observations are briefly described in Section II, while a discussion of the relationship between rotation and X-ray activity is presented in Section III.

II. Observations

Several observing runs at various ESO telescopes were devoted in 1983/84 to an extensive study of the Rho Ophiuchus dark cloud. This region was chosen because it is the only region of star formation which has already been observed repeatedly in the X-ray range by the "Einstein" satellite (Montmerle *et al.*, 1983). That many T Tauri stars found in this region have moderate emission characteristics also makes them suitable for a study of their rotation using spectroscopic techniques.

We present here the most recent results obtained in June 1984 at the 1.5 Danish telescope at La Silla using CORAVEL, an instrument generally used to measure radial velocities but also a very powerful tool for deriving stellar rotational velocities (cf. Benz and Mayor, 1981). CORAVEL correlates the incident stellar spectrum with a mask on which about 1,500 metallic lines are engraved. The position of the correlation peak gives the value of the radial velocity while the width of the peak measures the projected rotational velocity vsini. The detection limit is a few km/s.

This method was used to derive with good accuracy the rotational velocities of eight southern T Tauri stars. The rotational velocities of four additional T Tauri stars were measured using Fourier techniques on high resolution spectrograms obtained in February 1984 at La Silla at the 3.6 meter telescope with CASPEC. Combining these values with data found in the literature for 8 other T Tauri stars (Vogel and Kuhi, 1981 ; Rydgren *et al.*, 1984), as well as data for other late-type stars, we can now discuss the connection between rotation and activity in pre-main-sequence stars. In general, T Tauri stars prove much more active than even the most active main-sequence stars, so that several activity criteria can be defined and used in such a study. In this paper, we will restrict ourselves to a discussion of X-ray activity, but do intend to publish a more general discussion elsewhere.

III. Rotation and X-ray activity

We plot in Figure 1 the observed X-ray luminosity (suitably corrected for extinction, cf Montmerle et al., 1983) versus the projected rotational velocity for all late-type stars (G to M) for which such data exist. In Fig. 1, main-sequence late-type stars are represented by white symbols, RS CVn systems by star symbols and T Tauri stars by filled triangles. Bars associated with T Tauri stars represent the observed range of variability in X-ray luminosity as well as uncertainties on the rotational velocity. A least square fit of all data, drawn on Fig. 1, shows that the X-ray luminosity scales approximately as the square of the projected rotational velocity. Pallavicini et al., (1981) found the same observational relationship in their study of main sequence stars, and found this relation consistent with predictions from stellar dynamo models.

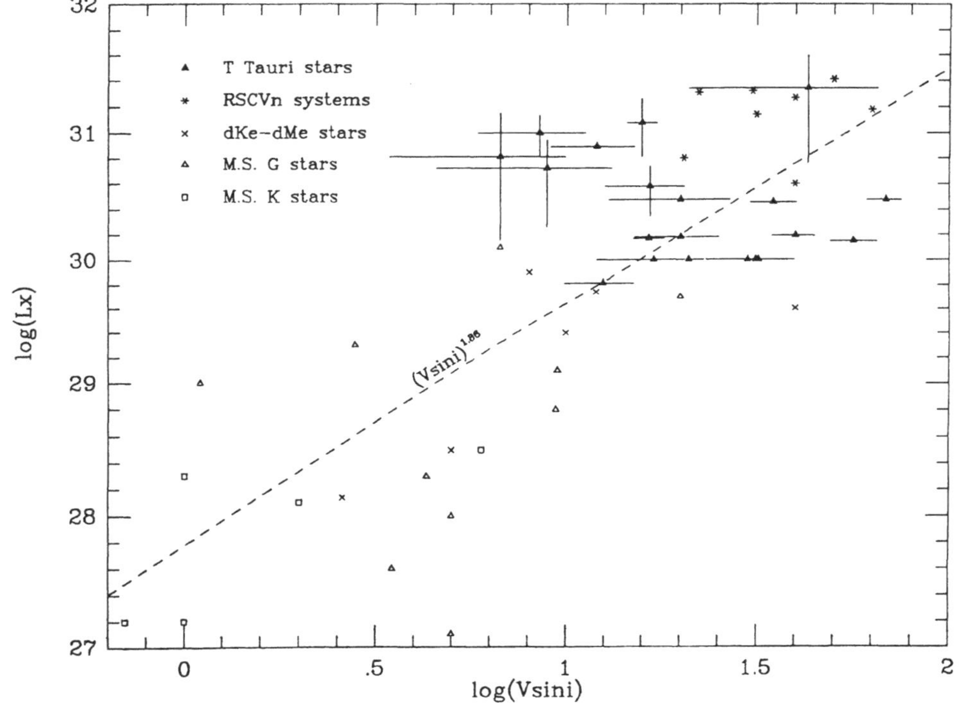

Fig.1 : Relationship between X-ray luminosity and rotation velocity for late-type stars.

From Figure 1, we conclude that the engine of X-ray activity is likely to be the same in T Tauri stars as in other late-type main-sequence and active stars. The enhanced activity of the T Tauri and RS CVn stars compared to main-sequence stars is accounted for simply by their higher rotational velocities.

RS Canis Venaticorum systems are late-type active, spectroscopic binaries. Their high velocities are due to the synchronization of their angular and orbital motions. Although they seem to lie higher than T Tauri stars in Fig. 1, this is due to a selection effect, since we only plot the most active systems for which we have accurate rotation velocities. X-ray surveys including RS CVn systems in fact show that, on the average, they have X-ray luminosities in the same range as those of T Tauri stars. The relationship discussed above between rotation and X-ray activity holds for the RS CVn class as a whole, not necessarily for its individual members, and the same is true for the T Tauri class. Since one of the main parameters of dynamo theories is the depth of the convection zone, this scatter can be attributed partly to the fact that we are dealing with different spectral types (from G to M). Also, the projected rotational velocity plotted in Fig. 1 only has statistical meaning since we expect rotation axises to be randomly distributed on the line of sight. We searched for but found no relationships between either rotation and age of the stars or rotation and Hα emission line-strength.

While the bulk of X-ray emission is of coronal origin in late-type main-sequence and active stars such as the RS CVn systems, the origin of X-ray emission in pre-main sequence stars is still controversial. On one hand, Walter and Kuhi (1981) found that T Tauri stars with strong Hα emission were not detected in the X-ray range. They concluded that coronae probably exist around all T Tauri stars, with varying degrees of "smothering" by absorption of X-rays in gaseous circumstellar envelopes. On the other hand, observations of the Rho Ophiochi region by Montmerle et al., (1983) demonstrated that flare-like, powerful X-ray eruptions are occurring in the atmosphere of many pre-main sequence stars of that region. Montmerle et al., propose that most of the X-ray emission in T Tauri stars is flare-like in nature, with possibly an additional contribution to the flux from a low emission-level, quiescent corona. That T Tauri stars follow the same relationship between X-ray flux level and rotation as main-sequence stars suggests a similar origin of at least part of the X-ray radiation, and is an argument for the presence of a quiescent corona with moderate emission measure and with X-ray flux close to or below the detection limit of "Einstein".

IV. Prospects

Work in progress includes a search for similar correlations between chromospheric activity indicators and rotation in the hope of better understanding what aspects of T Tauri activity can be attributed to magnetic phenomena. We also try to correlate these X-ray and chromospheric activity indicators with a parameter taking both the depth of the convective zone and the rotational velocity into account, in order to try to get a more accurate relationship between the strength of the different activity indicators and the efficiency of the dynamo process. This approach is needed if more precise statements are to be made about the dynamo process at work in T Tauri stars.

References

Benz W. and Mayor M., 1981, Astron. Astrophys., 93, 235
Montmerle T., et al., 1983, Ap.J., 269, 182
Noyes R.W., et al., 1984, Ap.J., 279, 763
Pallavicini R., et al., 1981, Ap.J., 248, 279
Rydgren A.E., et al., 1984, Astron. J., 89, 1015
Vogel S.N. and Kuhi L.V., 1981,Ap.J., 245, 960
Walter F.M. and Kuhi L.V., 1981, Ap.J., 250, 254

SUPERSONIC CO FLOWS, H_2O MASERS AND HH OBJECTS IN NGC 1333 AND NGC 7129

G. Sandell[(1)], R. Liseau[(2)]

(1) Observatory and Astrophys. Lab., Univ. of Helsinki, SF-00130 Helsinki, Finland
(2) Stockholm Observatory, S-13300 Saltsjöbaden, Sweden

ABSTRACT

We present high spatial resolution mapping (HPBW=32") of outflow regions in NGC 1333 and NGC 7129 in the J=1-0 CO and ^{13}CO lines. The blueshifted outflow related to the Herbig Haro objects HH 7-11 has been found to be clumpy and closely associated with the visible HH emission. The CO flow stops abruptly at HH 7, the outmost HH object of the group, as seen from the suspected excitation source, the low luminosity star SSV 13, also an H_2O maser. Although the flow is bipolar, the blue- and redshifted flows are not exactly aligned but inclined relative to each other at an angle $\sim 37^o$. We suggest that the redshifted flow is expanding through much denser gas, and that the flow is deflected by the surrounding dense cloud. South of HH 7-11, near the masers C and D, we have also found another outflow region, which we have not yet mapped in detail.

In the NGC 7129 region there are two distinct high velocity outflows, both of which appear to have their activity centers associated with H_2O masers and FIR sources. In the northern outflow, which probably is driven by the Herbig Be star LkH$_\alpha$ 234, most of the high velocity gas is observed to be redshifted, with very little gas in the opposite direction. The spatial distribution of ^{13}CO close to LkH$_\alpha$ 234 indicates strong density variations, which we interpret as a rotating gaseous disk, responsible for the collimation of the flow. The second outflow, associated with an deeply embedded star, has more symmetric wings.

1. INTRODUCTION

Herbig Haro objects are known to be cooling regions of high velocity schockwaves caused by strong stellar winds from embedded pre-main-sequence (PMS) stars (Schwartz, 1983). Sometimes HH objects appear to be related to well collimated optical jets (Mundt, 1984). In those few cases, where these PMS stars have been identified optically, they have been found to be T Tauri type stars, with the possible exception of R Mon (Jones and Herbig, 1982) and LkH$_\alpha$ 234 (this paper). The winds from these stars are also seen in the low velocity shocks producing molecular hydrogen emission (Zealey et al., 1984a) and the winds also drive the surrounding neutral gas, which is seen as high velocity molecular emission (Bally and Lada, 1983; Edwards and Snell, 1983;1984). These molecular outflows are rather well collimated and often appear as bipolar flows.

In a program to study the association between HH objects, molecular hydrogen emission, CO outflows and associated PMS objects, we have carried out detailed CO and ^{13}CO observations with the 20m telescope at Onsala Space Observatory, Sweden.

2. THE NGC 1333 REGION

The bright reflection nebula NGC 1333 in the dark cloud L 1450 is illuminated by the late B-type star BD+30°549. This dark cloud is part of the large Perseus OB2 association (Sargent, 1979). The region south of NGC 1333 is in a very active phase of PMS evolution. There are several known T Tauri stars and embedded IR sources (Strom et al., 1976). The catalogue by Herbig (1974) also lists 15 HH objects in this region. Recent surveys for H_2O masers and CO outflows have revealed at least five H_2O masers (Haschick et al., 1980;1983; Haschick and Henkel, personal comm.) and four separate CO outflows (Snell and Edwards, 1981; Edwards and Snell, 1983; Liseau and Sandell, 1983; Sandell et al., 1985). Herbig and Jones (1983) adopt a distance of 350pc to the dark cloud complex.

In the following we discuss more in detail the outflow associated with SSV 13, a star of relatively low mass, as judged from its luminosity, $\sim 45L_\odot$ (Harvey et al., 1984). This star is also one of the few low luminosity stars that coincides with an H_2O maser (Haschick et al., 1980). The star is obscured by ∿14^{m}5 of visual extinction and becomes visible only in the extreme red, where Liseau (1983) detected it as a 17^{m}1 star on an I band CCD image. The chain of HH objects, HH 7-11, points toward SSV 13, and all HH objects are highly blueshifted and have proper motions away from the star (Herbig and Jones, 1983). Mundt (1984) suggests that these HH objects outline an optical jet structure, with HH 7 representing the working surface of the jet and the rest of the HH objects being flow instabilities within the jet. The prominent bipolar CO outflow associated with HH 7-11 was first discovered by Snell and Edwards (1981).

We have mapped the blueshifted CO outflow with 15" resolution to allow a detailed comparison with the optical HH emission and with the schock excited molecular hydrogen at 2.1 μm mapped with 12" aperture by Zealey et al. (1984b). The redshifted outflow was mapped with a coarser grid. Due to bad weather we only got a few ^{13}CO spectra in the NGC 1333 region.

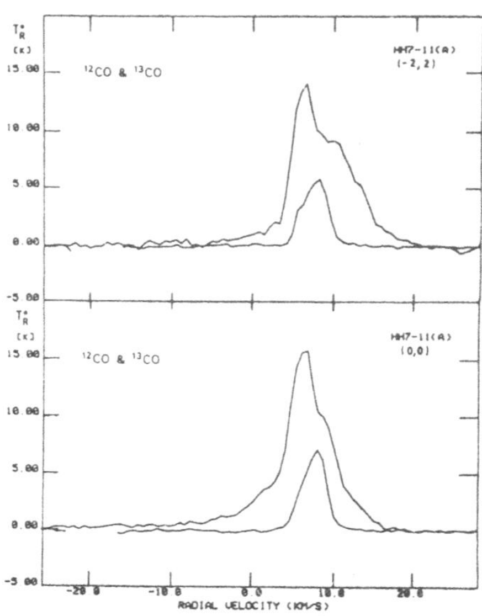

Fig. 1. CO and ^{13}CO spectra. (0,0) corresponds to SSV 13. Offsets are in units of 15 arcsec.

Fig. 1 shows CO and ^{13}CO spectra toward SSV 13 and a position 42" NW of the star. The blue wing is very prominent in the SSV 13 spectrum and extends in velocity over more than 35 km s^{-1}. The red wing is much less extended in velocity, with a maximum linewidth of ∿15 km s^{-1}, but is on the other hand stronger in intensity. Weak wings are also seen in ^{13}CO, but the blue flow is on the whole optically thin (c.f. Snell and Edwards, 1981). The ambient cloud emission has two blended components at velocities 6.3 and 8.3 km s^{-1}, which lead Loren (1976) to propose that cloud collision may have triggered star formation in the cloud. The 6 km s^{-1} component is rather diffuse and not seen in high excitation lines (see e.g. Lada et al., 1974; Schwartz et al., 1983). SSV 13 is most likely associated with the dense 8 km s^{-1} cloud, in which case the 6 km s^{-1} component is on the far side of the cloud, since the HH objects associated with the blue flow emerge at the cloud surface.

In Fig. 2 we show the integrated wing emission obtained by subtracting the ambient cloud emission from the spectra. Emission close to the line core was excluded from the line integral, since the ambient cloud emission was estimated from two-component Gaussian fittings, and is therefore uncertain near the line core. Although the red outflow is not very well sampled, the bipolar appearance of the outflow is quite clear. We also note that the blueshifted emission is closely associated with the visible HH emission. The close relation between these phenomena becomes even clearer if we deconvolve the map. In Fig. 3 we present a deconvolved map of the blue outflow obtained by removing a Gaussian beam of HPBW=32". Note that the flow appears to stop abruptly at

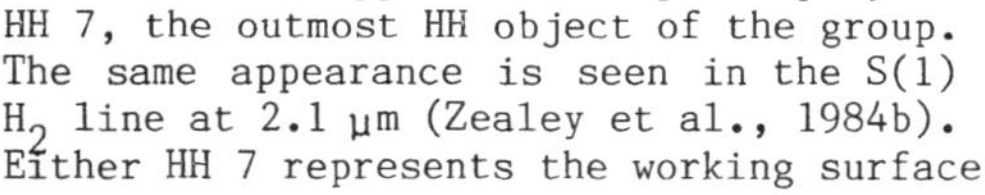

HH 7, the outmost HH object of the group. The same appearance is seen in the S(1) H_2 line at 2.1 μm (Zealey et al., 1984b). Either HH 7 represents the working surface

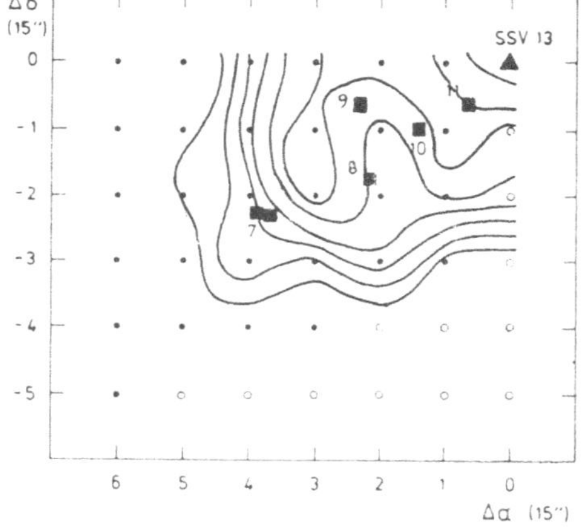

Fig. 3. A deconvolved map of the HH 7-11 blue flow. The intensity scale is arbitrary. Open circles denote positions of inter- or extrapolated intensity.

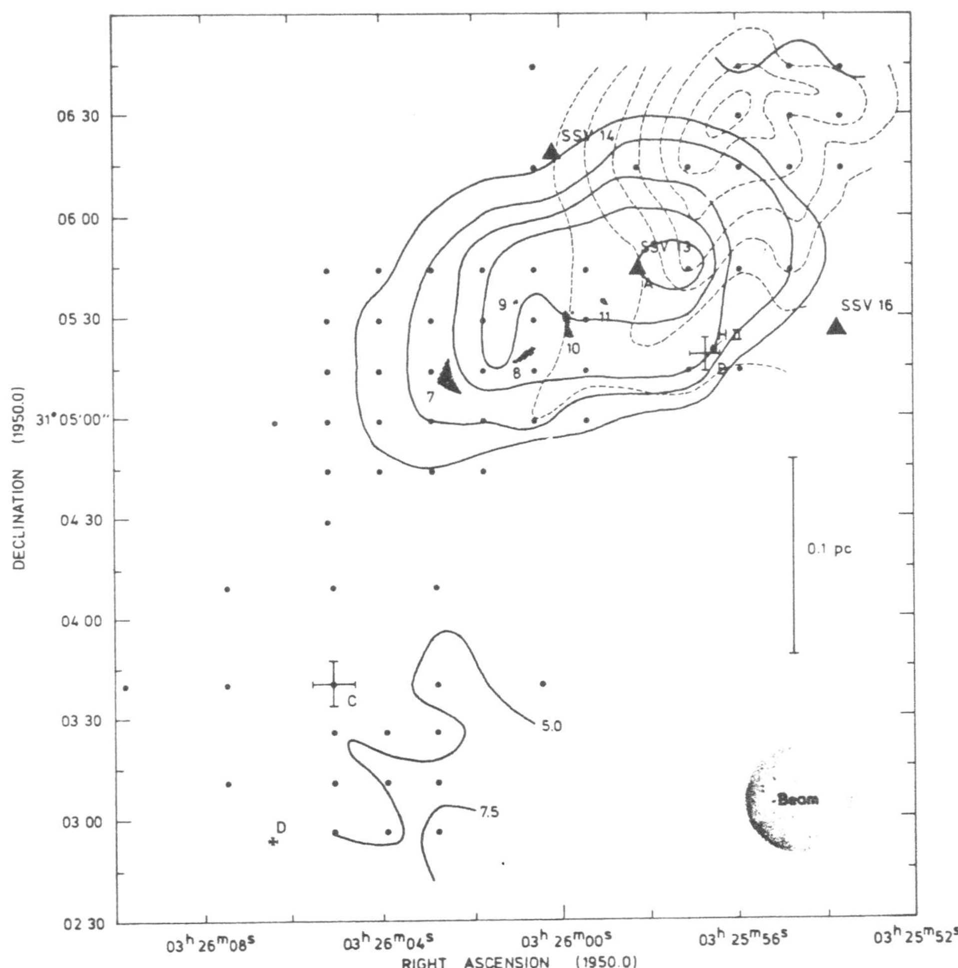

Fig. 2. A contour map of the integrated intensity in the blue (solid line) and red (dashed) CO wings. The isophotes start at a level of 5 K km s^{-1} and step in units of the same amount. The redshifted outflow near C lies below our lowest contour level. The blue wing is integrated over the velocity interval -30 to +3.6 km s^{-1}, while the red is between +10 and +25 km s^{-1}.

of the flow, or the CO stops at the surface of the cloud, where the wind has no more gas to interact with. The second hypothesis finds support from the timescales of the flow. The dynamical timescale of the blue flow is ∿4 times less than that of the red (Table 1). The blue outflow appears clumpy, and we see no evidence for limb brightening, which one would expect if the CO flow is dominantly formed by gas interacting with the wind only at the walls of a cavity formed by the wind (Canto, 1980).

It is also evident from Fig. 2, that the two lobes are not exactly aligned, but inclined by an angle of $\sim 37^{\circ}$. The simplest explanation for this is, that the red flow is deflected by a dense cloud wall, forcing the wind to expand in the direction where the gas density is lowest. The lower velocity of the red flow also supports the idea, that the redshifted flow is expanding into denser gas than the blueshifted one. Since SSV 13 is located on the frontside of a large cloud, this geometry is not evident from NH_3 and CS maps of the area (Ho and Barrett, 1980; Schwartz et al.,

1983). These do show a density ridge close to SSV 13, possibly a collimating disk (Schwartz et al., 1983), but the column density falls off in both directions. In this case the overall appearance of the cloud apparently masks the density variations close to the flow.

The total extent of the wings of the outflow tentatively associated with the C maser are only ~14 km s^{-1} compared to almost 50 for the SSV 13 case. No near or far IR source has yet been discovered near these masers, which suggests that the luminosity of the driving source is very low.

3. THE NGC 7129 REGION

The reflection nebula NGC 7129 is seen in projection onto the dark cloud L 1181. The nebula is illuminated by a group of early type stars, among them are two Herbig Ae/Be stars, viz. BD+65°1637 and LkH$_{\alpha}$ 234 (Herbig, 1960). Although this region is more distant, ~1 kpc (Racine, 1968), we also here find emission line

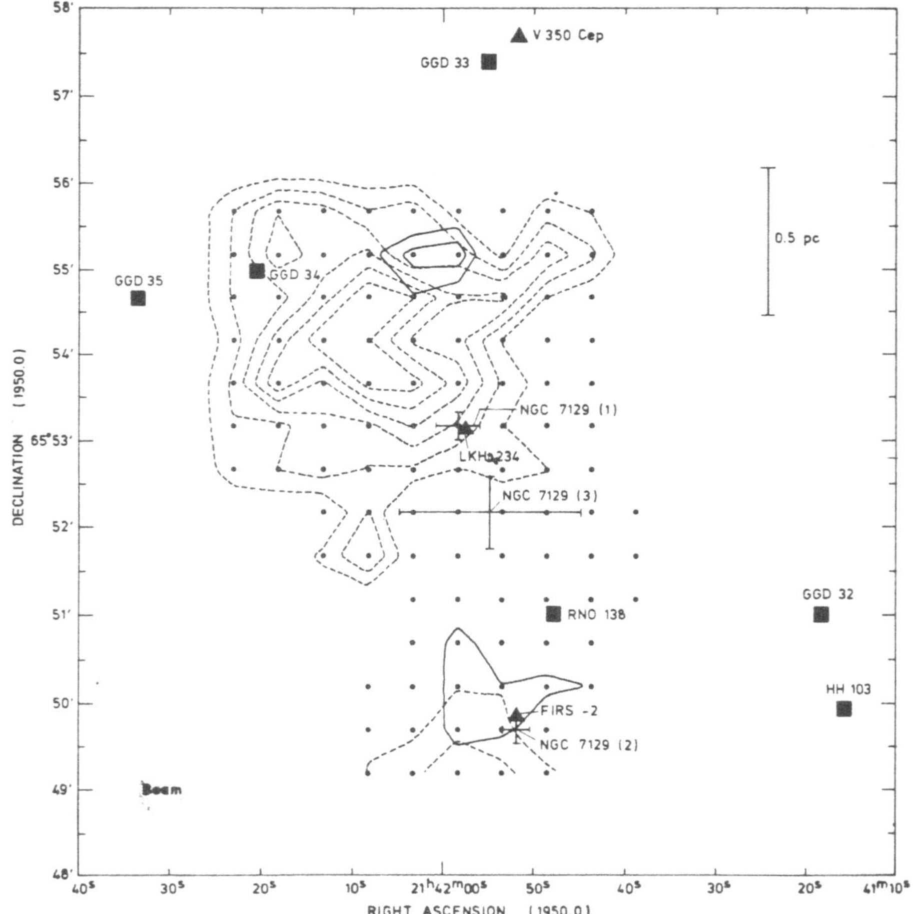

Fig. 4. The integrated wing emission of the outflows in NGC 7129. The blueshifted emission (solid lines) is integrated from -22 to -13 km s^{-1} with contour levels starting at .25 K km s^{-1}, while the redshifted contours (dashed) are integrated from -9 to +5 km s^{-1} and start from .4 K km s^{-1}. The contour levels are incremnts of these values. The filled squares designate red objects, triangles are IR sources and dots refer to observed positions.

stars, embedded IR sources and at least 3 spectroscopically identified HH objects: HH 103, GGD 32 and 35 (Strom et al., 1974; Magakyan, 1983; Cohen and Schwartz, 1983). Again we have additional evidence for active PMS evolution. There are three known H_2O masers (Sandell and Olofsson, 1981) and two spatially separated CO outflows (Liseau and Sandell, 1983; Edwards and Snell, 1983). One of the masers appears associated with LkH$_\alpha$ 234, the dominant FIR source ($L \sim 1200 L_o$) in the region (Bechis et al., 1978; Harvey et al., 1984). Recent VLA and VLBI observations (Rodriguez and Cantó, 1983; Sandell et al., 1984) resolve this maser into two components separated by $\sim$3", which are displaced from the star by $\sim$8" to NNW.

We have mapped a large area around LkH$_\alpha$ 234 with 30" spacing in both CO and ^{13}CO (Fig. 4 &5). The wings toward LkH$_\alpha$ 234 extend over more than 27 km s^{-1}, with the red wing dominating the spectrum. The red wing is optically thick, since the wing is also clearly seen in ^{13}CO. The kinetic temperature of the red outflow is probably below the temperature of the ambient gas, 16K, which corresponds to a beam filling factor of .2 for the high velocity gas. The red outflow peaks NE of LkH$_\alpha$ 234 and extends to GGD 34, which, however, apparently is not an HH object. The nearby GGD 35 is strongly blueshifted (Magakyan, 1983) and is therefore unrelated to the red outflow. Since LkH$_\alpha$ 234 is known to have strong winds (Finkenzeller and Mundt, 1983), and to be associated with H_2O maser emission, possibly even radio and X-ray emission (Bertout and Thum, 1982; Cassinelli, personal comm.), we tentatively identify it as the driving source of the flow. Both our CO and ^{13}CO maps show that LkH$_\alpha$ 234 lies close to the edge of the cloud. We find practically no emission SW of the star, which also explains the absence of blueshifted emission. Recent observations (Sandell et al., 1985) reveal, however, another cloud near GGD 32 and HH 103. Here the blueshifted wing emission reappears, now almost detached from the cloud core, indicating an acceleration of the flow, similar to what is seen in L 1551 (Fridlund et al., 1984). Therefore the LkH$_\alpha$ 234 outflow is also bipolar, with the red flow plowing into the dense cloud, while the blue flow becomes visible only close to the HH objects GGD 34 and HH 103, both of which have highly blueshifted emission lines (Strom et al., 1974; Magakyan, 1983). This is strikingly similar to R Mon/HH 39 case (Cantó et al., 1981).

The gas distribution around LkH$_\alpha$ 234 (Fig. 5) appears to outline a disk or toroid located perpendicular to the CO outflow. Presumably this disk collimates the CO flow. The diameter of the disk is $\sim$.45pc, and the thickness is at most one third of the diameter. Our ^{13}CO observations in the plane of the disk reveal a velocity gradient $dV/dr \sim 1$ km s^{-1} pc^{-1}, which corresponds to a rotational motion of $V\sin i \sim 0.15$ km s^{-1} at the outer edge of the disk. This disk appears to be somewhat larger and more slowly rotating than the recently discovered disks in L 1551 and NGC 2071 (Kaifu et al., 1984; Takano et al., 1984).

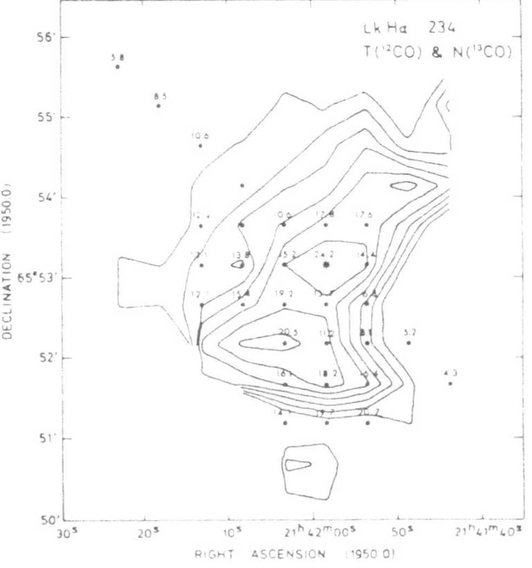

Fig. 5. A composite map of the distribution of CO radiation temperature and ^{13}CO column density around LkH$_\alpha$ 234. The CO temperature distribution is shown as a contour map starting at 11K with steps of 1K. The ^{13}CO column density is given in units of 10^{15}cm^{-2} and displayed at the observed postions (dots). At two positions the cloud emission was too heavily blended by wing emission, which prevented us from deriving reliable column densities. The asterisk marks the position of LkH$_\alpha$ 234.

TABLE 1
Physical characteristics of the outflows

Source	Wing	V_{max} [a] (km s^{-1})	R_{max} (pc)	t_d (10^3 yrs)	M [b] ($M_\odot$)	MV_{max} ($M_\odot$ km s^{-1})	$\dot{E}$ [c] ($L_\odot$)	L_* ($L_\odot$)	$\dot{M}$ [d] (3 $10^{-8} M_\odot$ yr^{-1})
SSV 13	blue	35	0.17	4.8	0.21	7.4	4.4	45	170
	red[e]	15	>0.18	>12.1	>0.15	>1.8	-		-
LkH 234	blue[e]	9	>0.33	>37	>0.08	>0.7	-	1200	-
	red	13	1.21	93	5.95	77.4	20		100
NGC 7129(2)	blue[e]	9	>0.46	>51	>0.17	>1.5	-	230	-
	red	8	0.58	72	0.44	3.5	0.03		5

Footnotes: [a] The maximum wing velocity relative to line center velocity.
[b] A beam filling factor of 0.3 is used for the blue SSV 13 outflow, and 0.2 for LkH_α 234. In all other cases the beam filling is taken to 1.
[c] E denotes the mechanical luminosity of the flow.
[d] A terminal wind velocity $V_\infty = 300$ km s^{-1} is assumed.
[e] Values are lower limits due to incomplete mapping.

4. DISCUSSION

Both regions studied show evidence for multiple CO outflows and H_2O maser emission, suggesting that an outflow phase is very common and that it occurs almost simultaneously in a star forming cloud. The outflows associated with SSV 13 and LkH_α 234 both appear to be strongly affected by the density distribution of the gas in the surrounding parent cloud. The high resolution maps of these outflows indicate that the CO gas is clumpy within these outflows and closely associated with the visible HH emission. The mass loss rates, that we derive from our CO observations (Table 1), are all much higher than the DeCampli limit for a $1M_\odot$ star (DeCampli, 1981), and suggest that the stars loose a large fraction of their mass, if the winds are continuous during the PMS stage. The dynamical timescales range from $\sim 10^4$ yrs for SSV 13 to $\sim 10^5$ yrs for the much more luminous star LkH_α 234. The CO outflows have collimation angles of 60° to 90° and may well be collimated by interstellar disks (Cantó, 1980; Königl, 1982). The extremely well collimated optical jets that are sometimes seen near T Tauri stars (Mundt, 1984) require, however, the collimation to occur much closer to the star.

Acknowledgements. The Onsala Space Observatory is operated by Chalmers University of Technology, Gothenburg, Sweden, with financial support from the Swedish Natural Research Council (NFR). We gratefully acknowledge financial support from NFR and from the Academy of Finland. We thank J. Högbom for help with the deconvolution algorithm, N. Holsti for help with the reduction software and B. Greder for the artwork. A more complete account of this work is to be submitted to Astron. Astrophys..

REFERENCES

Bally, J., Lada, C.J.: 1983 Astrophys. J. 265,824
Bechis, K.P., Harvey, P.M., Campbell, M.F., Hoffman, W.F.: 1978 Astrophys. J. 226, 438
Bertout, C., Thum, C.: 1982 Astron. Astrophys. 107, 368
Cantó, J.: 1980 Astron. Astrophys. 86, 327
Cantó, J., Rodríguez, L.F., Barral, J.F., Carral, P.: 1981 Astrophys. J. 244, 102
Cohen, M., Schwartz, R.D.: 1983 Astrophys. J. 265, 877
DeCampli, W.M.: 1981 Astrophys. J. 244, 124
Edwards, S., Snell, R.L.: 1983 Astrophys. J. 270, 605
Edwards, S., Snell, R.L.: 1984 Astrophys. J. 281, 237
Finkenzeller, U., Mundt, R.: 1984 Astron. Astrophys. Suppl. Ser. 55, 109

Fridlund, C.M.V., Sandqvist, Aa., Nordh, H.L., Olofsson, G.: 1984 Astron. Astrophys. 137, L17
Harvey, P.M., Wilking, B.A., Joy, M.: 1984 Astrophys. J. 278, 156
Haschick, A.D., Moran, J.M., Rodríguez, L.F., Burke, B.F., Greenfield, P., Garcia-Barreto, J.A.: 1980 Astrophys. J. 237, 26
Haschick, A.D., Moran, J.M., Rodríguez, L.F., Ho, P.T.P.: 1983 Astrophys. J. 265,281
Herbig, G.H.: 1960 Astrophys. J. Suppl. 4, 337
Herbig, G.H.: 1974 Lick Obs. Bulletin No. 658
Herbig, G.H., Jones, B.F.: 1983 Astron. J. 88, 1040
Ho, P.T.P., Barrett, A.H.: 1980, Astrophys. J. 237, 38
Jones, B.F., Herbig, G.H.: 1982 Astron. J. 87, 1223
Kaifu, N., Suzuki, S., Hasegawa, T., Morimoto, M., Inatani, J., Nagane, K.,Miyazawa, K., Chikada, Y., Kanzawa, T., Akabane, K.: 1984 Astron. Astrophys. 134, 7
Königl, A.: 1982 Astrophys. J. 261, 115
Lada, C.J., Gottlieb, C.A., Litvak, M.M., Lilley, A.E.: 1974 Astrophys. J. 194, 609
Liseau, R.: 1983 IAU Circular No. 3887
Liseau, R., Sandell, G.: 1983 Rev. Mexicana Astron. Astrof. 7, 199
Loren, R.B.: 1976 Astrophys. J. 209, 466
Magakyan, T.Yu.: 1983 Sov. Astron. Lett. 9, 83
Mundt, R.: 1983 to appear in Protostars and Planets II, Eds. D. Black and M. Mattews
Racine, R.: 1968, Astron. J. 73, 233
Rodríguez, L.F., Cantó, J.: 1983 Rev. Mexicana Astron. Astrof. 8, 163
Sandell, G., Olofsson. H.: 1981 Astron. Astrophys. 99, 80
Sandell, G. et al.: 1984 in preparation
Sandell, G., Liseau, R., Zealey, W.J., Williams, P.M.: 1985 in preparation
Sargent, A.I.: 1979 Astrophys. J. 233, 163
Schwartz, R.D.: 1983 Ann. Rev. Astron. Astrophys. 21, 209
Schwartz, P.R., Waak, J.A., Smith, H.A.: 1983 Astrophys. J. (Letters) 267, L109
Snell, R.L., Edwards, S.: 1981 Astrophys. J. 251, 103
Strom, S.E., Grasdalen, G.L., Strom, K.M.: 1974 Astrophys. J. 191, 111
Strom, S.E., Vrba, F.J., Strom, K.M.: 1976 Astron. J. 81, 314
Takano, T., Fukui, H., Ogawa, H., Takaba, R., Kawabe, Y., Fujimoto, K., Sugitani,K., Fujimoto, M.: 1984 Astrophys. J. (Letters) 282, L69
Zealey, W.J., Williams, P.M., Storey, J., Taylor, K., Sandell, G.: 1984a in Star Formation Workshop, Ed. R.D. Wolstencroft, Occasional Reports of the Royal Observatory, Edinburgh, No. 13, p. 109
Zealey, W.J., Williams, P.M., Sandell, G.: 1984b Astron. Astrophys. (in press)

VLA OBSERVATIONS OF THE OH AND H_2O MASERS IN THE YOUNG STAR CLUSTER NGC 2071 IR

G. Sandell[1)], L.Å. Nyman[2)], A. Haschick[3)] , A. Winnberg[2)]

1) Observatory and Astrophys. Lab., Univ. of Helsinki, Kopernikuksentie 1 SF-00130 Helsinki, Finland

2) Onsala Space Observatory, S-43900 Onsala, Sweden

3) Haystack Observatory and Harward-Smithsonian Center for Astrophysics, USA

ABSTRACT

We have mapped the maser emission in OH 205.1-14.1 (NGC 2071 IR), which is associated with a prominent bipolar outflow, at 1.6 and 22 GHz with the VLA in the A configuration. Our H_2O data are only preliminary, since we have checked only four velocity channels, but at these velocities the H_2O emission is resolved into at least three different point sources, each of which coincide with a compact HII region. At 1720 MHz we detected 100% polarized LC and RC maser emission from an unresolved point source. Since the LC and RC emission coincide in position, we interpret the line difference as due to Zeeman splitting, corresponding to a magnetic field of $\sim$14mG. The 1720 MHz maser has no clear radio or IR counterpart, but in the 6cm VLA continuum map, we see some faint emission, which could be due to ionized gas within the molecular flow, probably driven by IRS1, the strongest radio and H_2O source. The 6cm emission from IRS1 is extended and may outline a disk perpendicular to the flow. The 1667 MHz LC OH line was too weak to be detected.

1. INTRODUCTION

In the past, considerable effort has been made in understanding the nature of luminous H_2O maser sources, which are associated with O or early B stars and accompanied by class I OH masers. Lower luminosity maser sources, which are associated with stars of spectral class of middle B or later, show simple spectra and rapid intensity variations; they are more difficult to observe and thus less understood. In this paper we discuss the characteristics of the H_2O and OH maser OH 205.1-14.1, which is probably one of the lowest luminosity stars associated with a class I OH maser.

2. THE NGC 2071 REGION

The maser OH 205.1-14.1 was first discovered by Johansson et al. (1974). At that time it was only masering at the 1667 MHz transition and did not emit at any of the other OH ground state lines. Subsequent observations by Pankonin et al. (1977) revealed that this emission was left circularly polarized and strongly time variable. They also found time variable H_2O emission, first detected toward the OH maser by Schwartz and Buhl (1975). Pankonin et al. discovered, however, that the OH and the H_2O maser positions did not necessarily agree, and that the H_2O maser was emitted from two spatially separate components. Although weak 1665 MHz emission has been seen twice (Caswell et al., 1978; this paper), the uniqueness of this maser persists. The 1667 MHz OH line is always left circularly polarized and it has flared several times (Kazès et al., 1980; 1982). These flares appear to be correlated with H_2O flares, as seen from time monitoring of the H_2O maser (Sandell et al., 1985a). After the H_2O maser had its major flare (Mattila and Toriseva, 1981), it displayed a remarkable symmetry, indicative of a shell or rotating disk, lasting for about half a year (Lekht et al., 1982; Sandell et al., 1985a). Another symmetrical triple was seen in 1982. The 1720 MHz maser toward this region was first discovered by Kazès (personal comm.). It is also time variable and displays 100% circularly polarized emission.

The OH and H_2O maser are found toward a cluster of IR sources (Persson et al., 1981), with at least three, possibly four spatially separate components. Three of these are also seen as weak radio continuum sources (Snell and Bally, cited in Snell et al., 1984). The total luminosity of the cluster is $\sim 10^3 L_\odot$ (Harvey et al., 1979; Sargent et al., 1981) corresponding to stars of spectral class B3 or later. The dominant source is IRS1, which is also the strongest radio continuum source and has a spectrum consistent with that of an optically thin HII region (Simon et al., 1983). This star probably drives the prominent bipolar outflow, seen as high velocity molecular line emission and as shock excited H_2 emission (see e.g. Bally, 1982; Snell et al., 1984; Bally and Lane, 1982). Recent high resolution CS observations have revealed a compact disk perpendicular to the high velocity outflow (Takano et al., 1984). A deep CCD image taken by Reipurth on the 1.5m Danish telescope at La Scilla brings out a red nebulosity, possibly a Herbig Haro object.

3. OBSERVATIONS

Fig. 1 shows OH spectra obtained with the 100m Effelsberg telescope. Note especially the weak and very narrow 1665 MHz line ($\Delta V < 0.15$ km s^{-1}). The 1720 MHz lines agree in velocity with those observed during our VLA run. Fig. 2 shows an H_2O spectrum taken with the 37m telescope of the Haystack Observatory a few days before our VLA observations. Fig. 3 gives a 6cm VLA continuum map obtained by Winnberg and Turner. This map does not extend to IRS2, the weakest continuum source, for which Snell and Bally obtain a flux of only 0.5 mJy.

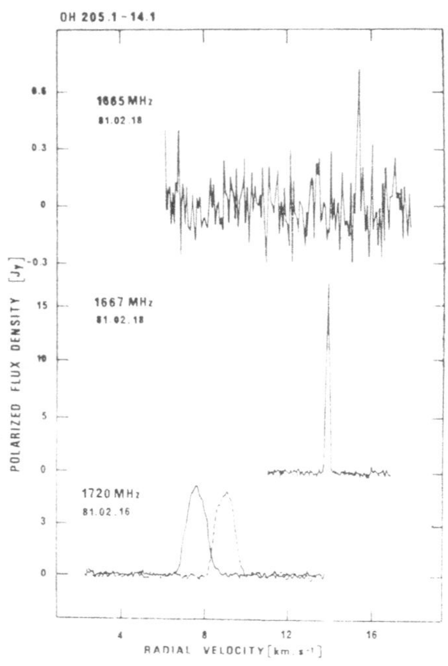

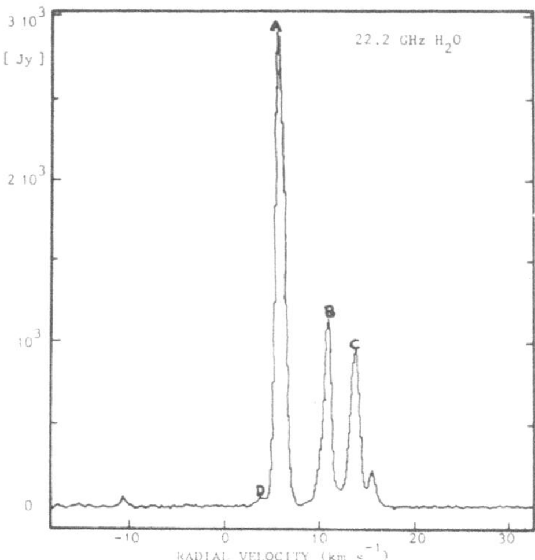

Fig. 2. Single dish H_2O spectrum. Only the VLA data corresponding to the components A-D have so far been processed.

Fig. 1. Single dish OH data. The main lines are observed in LC polarization. The dashed 1720 MHz line represents RC polarization.

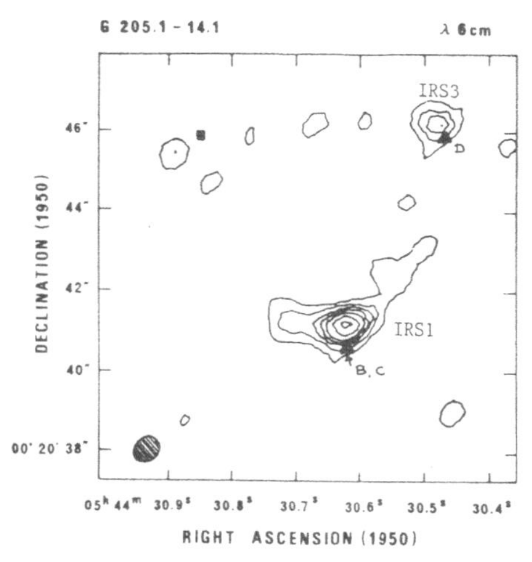

Fig. 3. A 6cm VLA continuum map. The triangles give the H_2O maser positions and the square the 1720 MHz OH maser. The integrated flux of IRS1 is ∿ 9 mJy and ∿4 mJy for IRS3. The extended emission around IRS1 is real and so is perhaps some of the faint emission northeast of the star.

VLA observations were carried out on Oct. 19 1983 with the A array, which gives maximum baselines of 35km. For the OH observations we used 25 antennas and 32 correlator channels (ΔV = 0.28 km s^{-1}) and for H_2O 18 antennas and 64 channels (ΔV = 0.66 km s^{-1}). The H_2O data are preliminary, since we have had time to reduce only 4 channels out of 64. The H_2O emission was very strong (c.f. Fig. 2), but because of atmospheric phase variations we were only able to deduce an absolute position for the strongest maser feature (at 6.4km s^{-1}). The rest of the data have been self calibrated using this line as a reference. We may have a slight position error in our absolute position (c.f. Fig. 3), but this can probably be removed by reprocessing the data. The 1720 MHz LC and RC emission are point sources compared to our 1"16 x 1"14 beam and coincide in position. The integrated fluxes at the line peak are 4.1 and 1.6 Jy for the RC and LC polarization, respectively. No 1667 MHz LC emission was seen to a limit of ∿.1 Jy. We have superposed the deduced OH and H_2O positions on the continuum map in Fig. 3. The strongest H_2O feature coincides with IRS2, which is outside the map.

4. DISCUSSION

Our high resolution H_2O observations show that the H_2O maser is emitted from at least three spatially different positions, all within 10". Since these coincide with compact HII regions, the masers are apparently excited by separate stars. Unfortunately the 1667 MHz OH line was too weak to be detected at the time of the observations, and we therefore do not know how it relates to the three H_2O masers. The best position estimate we have, obtained with the 100m Effelsberg telescope, gives a position α_{1950}= 05^h 44^m $32^s \pm 1^s$, δ_{1950}= 0° 20' 37" ± 15", i.e. closest to IRS2, but it could very well be any or none of the three. We find it significant, that all H_2O masers which are associated with low luminosity, high velocity flow regions, that so far has been measured with enough positional accuracy: HH7-11 (Haschick et al., 1980), NGC 2071 (this paper), all appear to coincide with stars, seen as IR and/or weak radio continuum sources. Therefore masers appear to be very close to their exciting stars if these are of low luminosity, as already proposed by Sandell and Olofsson (1981). A complete reduction of our H_2O data may able us to resolve some of the maser structure close to the star - outline the interaction with a disk or a jet - similar to what have been found by Cohen et al. (1984) in the case of the outflow associated with Ceph A. If the VLA resolution is not sufficient, the maser should at least be resolved by VLBI techniques. We have carried out a successful four station VLBI experiment, which, however, is not yet completely reduced (Sandell et al.,1985b). It should be noted, that our 6cm map does seem to outline a disk around IRS1, perpendicular to the molecular high velocity outflow.

The 1720 MHz maser does not have any obvious radio or IR counterpart, although there is some faint radio emission nearby (Fig. 3). Because the 1720 MHz line is blueshifted relative to the cloud and lies in the blue portion of the outflow (as seen from IRS1), it appears likely that it could be a cloudlet or a cloud ridge shocked by the wind. A very similar 1720 MHz OH maser has been found close to the FU Orionis star V1057 Cyg (Winnberg et al., 1981). Since we find that the left and right hand polarized maser emission coincide, the observed velocity difference appears to be due to Zeeman splitting, in which case the magnetic field is ∿ 14 mG. Another case, where 1720 MHz masers appear to outline a shock region, is seen in W28, where there is a string of masers along the SNR interaction zone (Wootten, 1981).

Acknowledgements: NRAO is operated by Associated Universities Inc. under contract with the National Science Foundation. We thank our friend at VLA, Ed Fomalont, and also Pat Palmer for help with the data reduction. We also thank Barry Turner for the 6cm data and Bo Reipurth for the CCD observations. This work has been funded by the Swedish Natural Science Research Council and the Academy of Finland.

REFERENCES

Bally, J.: 1982 Astrophys. J. 261, 558
Bally, J., Lane, A.P.: 1982 Astrophys. J. 257, 612
Caswell J.L. et al.: 1980 Austr. J. Phys. 33, 639 (see especially p. 667)
Cohen, R.J., Rowland, P.R., Blair, M.M.: 1984 Mon. Not. R. astr. Soc. 210, 425
Harvey, P.M., Campbell, M.F., Hoffmann, W.F., Thronson Jr., H.A.: 1979 Astrophys. J. 229, 990
Haschick, A.D., Moran, J.M., Rodríguez, L.F., Burke, B.F., Greenfield, P., Garcia-Barreto, J.A.: 1980 Astrophys. J. 237, 26
Johansson, L.E.B., Höglund, B., Winnberg, A., Nguyen-Q-Rieu, Goss, W.M.: 1974 Astrophys. J. 189, 455
Kazès, I., Cesarsky, D., Biraud, F., Drouhin, J.-P.: 1980 IAU Circ. No. 3502
Kazès, I., Biraud, F., Drouhin, J.-P.: 1982 IAU Circ. No. 3700
Lekht, E.E., Pashchenko, G.M., Rudnitskiî, G.M., Sorochenko, R.L.: 1982 Sov. Astron. 26, 168
Mattila, K., Toriseva, M.: 1981 IAU Circ. No. 3570
Persson, S.E., Geballe, T.R., Simon, T., Lonsdale, C.J., Baas, B.F.: Astrophys. J. (Letters) 251, L85
Sandell, G., Olofsson, H.: 1981 Astron. Astrophys. 99, 80
Sandell, G., et al.: 1985a in preparation
Sandell, G., et al.: 1985b in preparation
Sargent, A.I., van Duinen, R.J., Fridlund, C.V.M., Nordh, H.L., Aalders, J.W.G.: 1981 Astrophys. J. 249, 607
Simon, M., Felli, M., Cassar, L., Fischer, J., Massi, M.: 1983 Astrophys. J. 266,623
Snell, R.L., Scoville,N.Z., Sanders,D.B., Ericksson,N.R.: 1984 Astrophys. J. 284,176
Takano, T., Fukuy, Y., Ogawa, H., Takaba, H., Kawabe, R., Fujimoto, Y., Sugitani, K. , Fujimoto, M.: 1984 Astrophys. J. (Letters) 282, L69
Winnberg, A., Graham, D., Walmsley, C.M., Booth, R.S.: 1981 Astron. Astrophys. 93,79
Wootten, A.: 1981 Astrophys. J. 245, 105

TWO ZAMS AOV STARS IN CHA I

P.R. Wesselius[1), P.S. The[2), H.R.E. Tjin A Djie[2), H. Steenman[2).

1) Laboratory for Space Research,
P.O. Box 800, 9700 AV Groningen, The Netherlands.

2) Astronomical Institute 'Anton Pannekoek', University of Amsterdam,
Roetersstraat 15, 1018 WB Amsterdam, The Netherlands.

This report is based in part on observations collected at the European Southern Observatory La Silla, Chile, and on IRAS observations. The Infrared Astronomical Satellite was developed and operated by The Netherlands Agency for Aerospace Programs (NIVR), the US National Aeronautics and Space Administration (NASA), and the UK Science and Engineering Research Council (SERC).

1. INTRODUCTION

We discuss the balance between absorbed and re-emitted flux for pre-main-sequence stars surrounded by circumstellar dust, in particular for the stars HD 97048 and HD 97300 in the Cha I association. The MK spectral classification and apparent visual magnitude for these stars are: B9.5 Ve and V = 8.4; B9V and V = 9.0, respectively. Both objects are presumably ZAMS stars; although HD97300 has no emission features, it does have many of the properties of a Herbig Ae/Be star.

JHKLM observations for HD97048 were interpreted by Glass (1979) as due to a dust shell of 800 K while similar photometry for HD97300 was explained by Hyland et al (1982) in terms of a stellar companion of 3000 K. Our ground-based observations have shown that both stars are not variable. HD 97048 has exceptional emission features at 3.4 and 3.5 μm (Blades and Whittet, 1980).

2. OBSERVATIONS AND EXTINCTION/ABSORPTION CORRECTION

A large amount of new data on both stars has been assembled, consisting of low-dispersion IUE spectra from the VILSPA Data Banks, IDS and Reticon spectra as well as photometry in the Walraven, Johnson, Stromgren and Cousins systems at ESO, and IRAS far-infrared photometry (a more extensive discussion of the IRAS observations will be published separately). In the two figures all data are summarized: the fluxes were all put on the same absolute energy scale using the different absolute calibrations. This results in a systematic uncertainty from point to point of 10 to 20 %, at least beyond ~0.8 μ. The M and N values have additional large observational errors and may be uncertain to 40 %.

Because we want to compare the energy absorbed and reemitted by circumstellar dust we first have to correct for interstellar extinction. The F5V star HD 97240 is situated close to the direction of HD 97048 and has an E(B-V) of 0.05. Fitting the observed energy distribution in the BVRI part to the appropriate Kurucz (1979) model we find R = 3.1. From this value of R and from the fact that HD 97240 has no infrared excess we infer that its extinction is only due to foreground interstellar dust. The "observed" fluxes in the two figures have been corrected for an interstellar extinction of E(B-V) of 0.1 using the extinction law of Savage and Mathis (1979). (Below it will be derived that Cha I lies at ≈ 200 pc.)

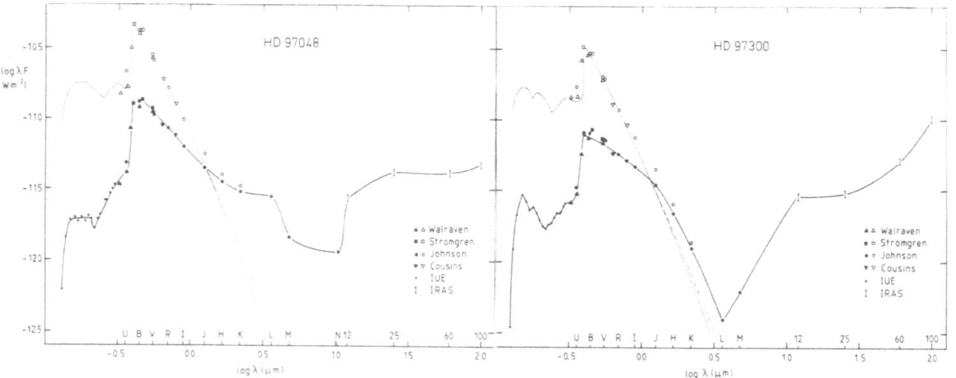

The flux originally emitted from the stellar object has been estimated under two basic assumptions: the circumstellar dust does not affect the stellar radiation in the spectral region around the J band (1.25 μ) and a Kurucz(1979) model of the appropriate temperature accurately describes the energy distributions for these stars. The Kurucz model was fitted at J and is shown as a thin line in the figures. R appears to be ~ 3.1 by fitting the "observed" energy distribution to the Kurucz model in the range B to I, assuming the interstellar extinction law applies to circumstellar absorption in this range. Based upon the spectral types of our program stars, with R = 3, and the absolute visual magnitude according to Schmidt-Kaler (1982) the distance of the Cha I cloud is determined to be ~ 200 pc.

This approach to obtain the extinction-free energy distribution (especially necessary for the ultraviolet) is supported by the conventional correction of the "observed" points around the Balmer jump. The resulting corrected points are fitted well by the Kurucz curve. In the ultraviolet we find a rather strange circumstellar absorption curve. This is not surprising because even the shape of the interstellar extinction curve varies from region to region by large amounts (Wesselius, 1982).

3. ENERGY BALANCE

Our calculations show that the possible contribution of free-bound and free-free emission to the huge infrared excess is negligible (dotted curve). Thus the absorption in the ultraviolet and visual on the one hand and the huge infrared excess on the other hand are presumably caused by the circumstellar dust of the Cha I complex. Then, we would expect an energy balance. However, we find:

Star	Energy absorbed	Energy reemitted
HD 97048	3	1
HD 97300	2.3	1.3

The CPC map for HD 97300 discussed by Wesselius et al (1984) shows that the infrared energy is emitted by two equally strong sources, HD97300 and H23/H24. Thus the reemitted energy of HD97300 alone is ~ 0.7. (All energies are in units of 10^{-11} Wm^{-2}.)

Surprisingly enough, only one third of the energy absorbed in the visual and ultraviolet is reemitted between 1 and 100 μm. There are two ways to account for the "missing" energy. At wavelengths longer than 100 μm a considerable amount of energy may be emitted, the size of the emitting region increases with wavelength (already a few arc minutes at 100 μm, see the CPC maps of Wesselius et al, 1984). The total energy emitted by the surrounding reflection nebula of each star is originally due to the illuminating star. We did not yet try to estimate the size of each contribution.

4. REFERENCES

Blades, J.C., Whittet, D.C.B. 1980, MNRAS 191 , 701. Glass, I.S. 1979, MNRAS 187 , 305.
Hyland, A.R., Jones, T.J., Mitchell, R.M. 1982, MNRAS 201 , 1095.
Kurucz, R.L. 1979, Astrophys. J. Suppl. Ser. 40 ,1.
Savage, B.D., Mathis, J.S. 1979, Ann. Rev. Astr. Ap. 17 , 73.
Schmidt-Kaler, Th. 1982, Landolt Börnstein Tables, Springer Verlag, 2 , 18.
Wesselius, P.R. 1982, ESA SP-182, p. 15.
Wesselius, P.R., Beintema, D.A., Olnon, F.M. 1984, Astrophys. J. 278 , L37.

Lecture Notes in Physics

Vol. 214: H. Moraal, Classical, Discrete Spin Models. VII, 251 pages. 1984.

Vol. 215: Computing in Accelerator Design and Operation. Proceedings, 1983. Edited by W. Busse and R. Zelazny. XII, 574 pages. 1984.

Vol. 216: Applications of Field Theory to Statistical Mechanics. Proceedings, 1984. Edited by L. Garrido. VIII, 352 pages. 1985.

Vol. 217: Charge Density Waves in Solids. Proceedings, 1984. Edited by Gy. Hutiray and J. Sólyom. XIV, 541 pages. 1985.

Vol. 218: Ninth International Conference on Numerical Methods in Fluid Dynamics. Edited by Soubbaramayer and J. P. Boujot. X, 612 pages. 1985.

Vol. 219: Fusion Reactions Below the Coulomb Barrier. Proceedings, 1984. Edited by S. G. Steadman. VII, 351 pages. 1985.

Vol. 220: W. Dittrich, M. Reuter, Effective Lagrangians in Quantum Electrodynamics. V, 244 pages. 1985.

Vol. 221: Quark Matter '84. Proceedings, 1984. Edited by K. Kajantie. VI, 305 pages. 1985.

Vol. 222: A. García, P. Kielanowski, The Beta Decay of Hyperons. Edited by A. Bohm. VIII, 173 pages. 1985.

Vol. 223: H. Saller, Vereinheitlichte Feldtheorien der Elementarteilchen. IX, 157 Seiten. 1985.

Vol. 224: Supernovae as Distance Indicators. Proceedings, 1984. Edited by N. Bartel. VI, 226 pages. 1985.

Vol. 225: B. Müller, The Physics of the Quark-Gluon Plasma. VII, 142 pages. 1985.

Vol. 226: Non-Linear Equations in Classical and Quantum Field Theory. Proceedings, 1983/84. Edited by N. Sanchez. VII, 400 pages. 1985.

Vol. 227: J.-P. Eckmann, P. Wittwer, Computer Methods and Borel Summability Applied to Feigenbaum's Equation. XIV, 297 pages. 1985.

Vol. 228: Thermodynamics and Constitutive Equations. Proceedings, 1982. Edited by G. Grioli. V, 257 pages. 1985.

Vol. 229: Fundamentals of Laser Interactions. Proceedings, 1985. Edited by F. Ehlotzky. IX, 314 pages. 1985.

Vol. 230: Macroscopic Modelling of Turbulent FLows. Proceedings, 1984. Edited by U. Frisch, J. B. Keller, G. Papanicolaou and O. Pironneau. X, 360 pages. 1985.

Vol. 231: Hadrons and Heavy Ions. Proceedings, 1984. Edited by W. D. Heiss. VII, 458 pages. 1985.

Vol. 232: New Aspects of Galaxy Photometry. Proceedings, 1984. Edited by J.-L. Nieto. XIII, 350 pages. 1985.

Vol. 233: High Resolution in Solar Physics. Proceedings, 1984. Edited by R. Muller. VII, 320 pages. 1985.

Vol. 234: Electron and Photon Interactions at Intermediate Energies. Proceedings, 1984. Edited by D. Menze, W. Pfeil and W. J. Schwille. VII, 481 pages. 1985.

Vol. 235: G. E. A. Meier, F. Obermeier (Eds.), Flow of Real Fluids. VIII, 348 pages. 1985.

Vol. 236: Advanced Methods in the Evaluation of Nuclear Scattering Data. Proceedings, 1985. Edited by H. J. Krappe and R. Lipperheide. VI, 364 pages. 1985.

Vol. 237: Nearby Molecular Clouds. Proceedings, 1984. Edited by G. Serra. IX, 242 pages. 1985.

Selected Issues from

Lecture Notes in Mathematics